Perspectives and Insights on Soil Contamination and Effective Remediation Techniques

Edited by Khalid Rehman Hakeem

Published in London, United Kingdom

Perspectives and Insights on Soil Contamination and Effective Remediation Techniques
http://dx.doi.org/10.5772/10.5772/intechopen.1003336
Edited by Khalid Rehman Hakeem

Contributors
Albert Kobina Mensah, Amir Khoshgoftar, Emmanuel Amoakwah, José G. Chan-Quijano, Karla L. Torres-López, Mahdi Khodaparast, Mary Allagoa, Mojtaba Khayam Nekouei, Ricardo Rodríguez-Cabrera, Shahram Naeimi, Soheila Aghaei Dargiri

First published in London, United Kingdom, 2024 by IntechOpen
IntechOpen is the global imprint of INTECHOPEN LIMITED, registered in England and Wales, registration number: 11086078, 167-169 Great Portland Street, London, W1W 5PF, United Kingdom

British Library Cataloguing-in-Publication Data
A catalogue record for this book is available from the British Library

Additional hard and PDF copies can be obtained from orders@intechopen.com

Perspectives and Insights on Soil Contamination and Effective Remediation Techniques
Edited by Khalid Rehman Hakeem
p. cm.

This title is part of the Environmental Sciences Book Series, Volume 22
Topic: Pollution
Series Editor: J. Kevin Summers
Topic Editors: Ismail Md. Mofizur Rahman and Zinnat Ara Begum

Print ISBN 978-0-85466-293-7
Online ISBN 978-0-85466-292-0
eBook (PDF) ISBN 978-0-85466-294-4
ISSN 2754-6713

For EU product safety concerns:
IN TECH d.o.o., Prolaz Marije Krucifikse Kozulić 3, 51000 Rijeka, Croatia,
info@intechopen.com or visit our website at intechopen.com.

IntechOpen Book Series
Environmental Sciences
Volume 22

Aims and Scope of the Series

Scientists have long researched to understand the environment and man's place in it. The search for this knowledge grows in importance as rapid increases in population and economic development intensify humans' stresses on ecosystems. Fortunately, rapid increases in multiple scientific areas are advancing our understanding of environmental sciences. Breakthroughs in computing, molecular biology, ecology, and sustainability science are enhancing our ability to utilize environmental sciences to address real-world problems.

The four topics of this book series - Pollution; Environmental Resilience and Management; Ecosystems and Biodiversity; and Water Science - will address important areas of advancement in the environmental sciences. They will represent an excellent initial grouping of published works on these critical topics.

Meet the Series Editor

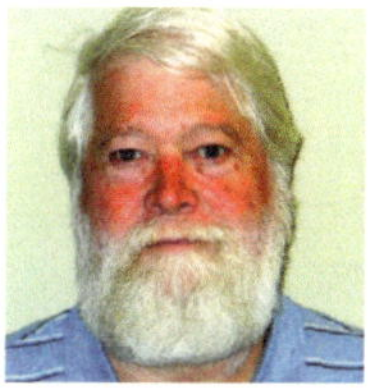

J. Kevin Summers is a Senior Research Ecologist at the Environmental Protection Agency's (EPA) Gulf Ecosystem Measurement and Modeling Division. He is currently working with colleagues in the Sustainable and Healthy Communities Program to develop an index of community resilience to natural hazards, an index of human well-being that can be linked to changes in the ecosystem, social and economic services, and a community sustainability tool for communities with populations under 40,000. He leads research efforts for indicator and indices development. Dr. Summers is a systems ecologist and began his career at the EPA in 1989 and has worked in various programs and capacities. This includes leading the National Coastal Assessment in collaboration with the Office of Water which culminated in the award-winning National Coastal Condition Report series (four volumes between 2001 and 2012), and which integrates water quality, sediment quality, habitat, and biological data to assess the ecosystem condition of the United States estuaries. He was acting National Program Director for Ecology for the EPA between 2004 and 2006. He has authored approximately 150 peer-reviewed journal articles, book chapters, and reports and has received many awards for technical accomplishments from the EPA and from outside of the agency. Dr. Summers holds a BA in Zoology and Psychology, an MA in Ecology, and Ph.D. in Systems Ecology/Biology.

Meet the Volume Editor

Dr. Khalid R. Hakeem is a professor at King Abdulaziz University, Jeddah, Saudi Arabia. He is currently the head of the Plant Biotechnology Research Unit, Princess Dr. Najlaa Bint Saud Al Saud Center for Distinguished Research in Biotech at the same university. He is also a visiting professor at Daffodil International University, Dhaka, Bangladesh; an adjunct research faculty at Chandigarh University, Panjab India and Chitkara University, Punjab; and a research fellow at INTI International University, Malaysia. Dr. Khalid has more than 14 years of teaching and research experience in plant ecophysiology and biotechnology. He is the recipient of several national and international fellowships. To date, Dr. Khalid has authored and edited more than 100 books with international publishers. He has also published 200 research articles in peer-reviewed international journals and 80 book chapters. Currently, Prof. Khalid is engaged in studying plant processes at ecophysiological and molecular levels.

Contents

Preface

Pollution resulting from anthropogenic actions represents a serious threat to life on Earth, and to restore damaged ecosystems, it is necessary to implement remediation strategies. These strategies can involve living organisms (biological remediation) or nonliving sources (physicochemical remediation). However, biological strategies are often more cost-effective for large-scale remediation of organic and inorganic pollutants.

Addressing soil contamination requires a multifaceted approach that combines monitoring, assessment, and targeted remediation techniques. By considering different perspectives and insights on this issue, we can work towards protecting the environment, promoting sustainability, and safeguarding human health.

This volume includes six chapters written by experts that describe soil contamination sources and remediation strategies. The book provides an overview of ecosystem approaches and phytotechnologies and their cumulative significance in solving various environmental problems. This book's main purpose is to offer readers, ranging from undergraduate students to researchers, extensive knowledge of soil contamination and the various strategies proposed to remediate these pollutants.

I am highly grateful to all our contributors for readily accepting our invitation to not only share their knowledge and research but also for unifying their diverse knowledge and expertise to create their chapters. I greatly appreciate their commitment.

I am also grateful to the staff at IntechOpen, particularly Mr. Josip Knapic, for their generous cooperation at every stage of the book's production.

Dr. Khalid Rehman Hakeem
Faculty of Science,
Department of Biological Sciences,
King Abdulaziz University,
Jeddah, Saudi Arabia

University Centre for Research and Development (UCRD),
Chandigarh University,
Punjab, India

Chapter 1

Effect of Contamination on Geotechnical Characteristics of Soil

Mahdi Khodaparast and Amir Khoshgoftar

Abstract

Soil contamination by petroleum contaminants and their derivatives considers one of the most important environmental hazards all around the world. Nowadays, with development of oil industry, there is a high probability of oil spills into the soil. This phenomenon, in addition to the environmental damages, causes changes in the physical, chemical and mechanical properties of soil. The effects of pollutants on granular and fine-grained soils will differ. This effect is physical in granular soil and changes the soil texture and structure in cohesive fine-grained soil. This chapter reviews the results of adding petroleum contaminants to the soil. it showed that changes in the geotechnical parameters of the soils were dependent on the type of soil, type of pollutant and duration of contamination. The primary effects were decreases in the internal friction angle, California bearing ratio and permeability of the soil and increases in the cohesion and Atterberg limits of the soil. The shear strength of the contaminated soil showed no definite or constant trend of change. When contaminated with acidic sludge (burnt-oil waste derivatives), despite an increase in the cohesion of the soil, a decrease in the internal friction angle caused a decrease in the shear strength.

Keywords: direct shear test, shear strength, petroleum pollutants, geotechnical properties, soil structure

1. Introduction

The effect of hydrocarbon pollution on the soil has attracted attention in recent years [1, 2]. Such pollutants not only damage the soil ecosystem and change its physical and mechanical properties, but also endanger the safety of civil engineering structures [3]. Petroleum is used to produce valuable products such as plastic, chemical fertilizers and as raw materials for the production of chemicals. More than two million tons of oil are produced globally every day, and about 10% of petroleum and its derivatives seep into underground water from leakages from storage tanks, pipelines, tankers and as waste from factories. In addition to reaching underground water, the pollution will spread horizontally through the soil by capillary forces to pollute even more soil [4–6].

Studies conducted in 1978–1992 have reported 33% of water and soil pollution resulted from improper storage and transportation and 67% resulted from defects and accidents during transportation of oil products via pipelines and oil tankers [7]. Although washing, as well as chemical and biological methods, have been introduced to clean contaminated soil, they are costly and have limitations [8]. An in-depth study of the geotechnical behavior of contaminated soil indicates that it would be more economical to use contaminated soil onsite for road infrastructure or other construction projects [9, 10].

Crude oil and its derivatives are mainly composed of hydrocarbons, nitrogen compounds, sulfur, organic metal compounds and mineral salts [11]. The duration of contamination, type of pollutant and the type of soil will affect the geotechnical properties of the polluted soil [12]. Any change in the shear resistance and other engineering characteristics of contaminated soil layers can cause a decrease in the stability of slopes, decrease the bearing capacity and increase the total and relative settlement of structural foundations [13].

The penetration rate of oil pollutants into the soil will depend on the soil characteristics (porosity, permeability, water content, etc.), as well as the nature and quantity of compounds in the pollutant [14, 15]. Fine-grained soils have a higher probability of contamination with petroleum substances and their derivatives because they have a larger specific surface area than coarse-grained soils. Fine et al. [14] introduced a parameter called the sensitivity index ranging from 0 to 1 for different types of soil. This index for sand should fall into the range of 0.01–0.1 and for clay in the range of 0.6–0.9.

Pollution in coarse-grained soil will cause physical interactions where the pollutant is connected to the soil by a weak van der Waals bond. In fine-grained soil, it will cause physical and chemical interactions as the pollutants are generally connected by covalent bonds. In clay soil, it will cause a decrease in the double layer thickness and a decrease in the dielectric constant, the result of which will be flocculation of the clay soil [16]. In order to improve its geotechnical characteristics, materials such as cement, lime, fly ash and nano-additives have been added [17, 18].

The soil investigated in this chapter study has exposed to pollutants and waste from a petroleum factory. This included the waste from burnt oil from a refinery. Different studies have been conducted on the effect of hydrocarbons and petroleum pollutants, including crude oil, gasoline, diesel, motor oil and burnt motor oil in fine and coarse soils, but little study has been done on soil contamination by waste from burnt oil from refineries [19–21]. Large-scale oil refinery fires have occurred in Salafchegan industrial region of Qom province in Iran and many other countries. The resulting waste products spread into the ground, causing extensive soil contamination. Dirt filter contaminants and acidic sludge are primary types of burnt oil refinery waste and the effects of these on the strength and compaction parameters of clayey sand are discussed in this chapter. The mechanism by which acidic sludge affects soil properties has been investigated and indicates that acidic sludge plays a determining role in the soil quality.

2. Oil contaminants and derivatives

Over the last few decades there has been an increased public awareness of environmental issues, particularly when the contamination of sand, water, and air is involved. Worldwide, scientists and environmentalists are faced with the challenge of overcoming the detrimental effects of the contamination of sand, air, and water. Oil leakage from pipelines, underground and surface fuel storage tanks, pollution by oil derivatives, especially waste from burnt oil from a refinery, waste disposal and management

Acid sludge	L.O.I	Na_2O	MgO	Al_2O_3	SiO_2	P_2O_5	SO_3	Cl	K_2O	CaO	Cr	Mo
	86.87	0.425	0.09	0.079	0.322	1.207	6.683	0.008	0.099	2.941	0.01	0.019
	Mn	Fe_2O_3	Ni	Cu	Zn	Sr	Zr	Pb				
	0.007	0.428	00.04	0.018	0.784	0.002	0.001	0.004				
Dirt filter	L.O.I	Na_2O	MgO	Al_2O_3	SiO_2	P_2O_5	SO_3	Cl	K_2O	CaO	TiO_2	Mn
	37.25	0.073	1.44	6.67	47.734	0.04	2.418	0.073	0.358	3.041	0.067	0.028
	Fe_2O_3	Ni	Zn	Rb	Sr	Zr	Ba					
	0.736	0.004	0.007	0.002	0.016	0.005	0.04					

Table 1.
XRF test results on contaminants.

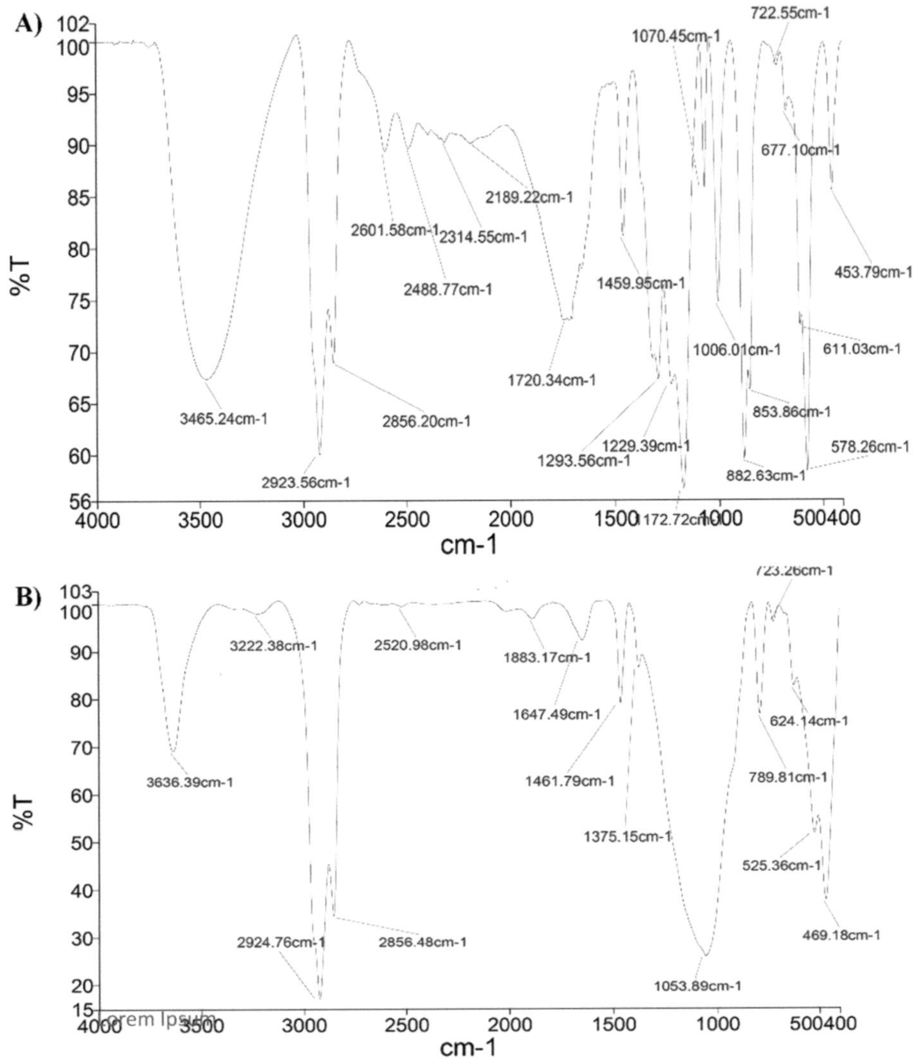

Figure 1.
FTIR results for: (a) acidic sludge; (b) dirt filter contaminants.

are the main sources of oil pollution. In this chapter firstly, different types of soil with different weight percentages were mixed with crude oil and its derivatives and then tested to determine the effect of the pollutant on the shear strength, compaction, Atterberg limits, permeability, etc. of the soil and secondly, to investigate the effect of soil pollution caused by different percentages of acidic sludge and dirt filter residue resulting from the treatment of burnt oil (0%, 3%, 6%, 9%), compaction tests were done on the contaminated soil samples according to ASTM D698 [22].

The primary components of these two contaminants as determined from x-ray florescence (XRF) and Fourier-transform infrared spectroscopy (FTIR) tests were ferrum, cuprum, plumbum, zinc, phosphorus, dust, halogenated compounds, glycols, carbon, polymers, asphaltene, polymethacrylates, phenols, sulfonates, amines, phosphonates, and thiophosphonates. The results of XRF and FTIR testing are shown in **Table 1** and **Figure 1** respectively.

3. Effect of contamination on soil geotechnical parameters

3.1 Effect of contamination on soil compaction parameters

In order to investigate the effect of burnt oil refinery waste on the maximum dry unit weight of the soil, modified compaction tests were performed at zero moisture content for different percentages of contaminants. **Figure 2** shows that, generally, with the increase of the polluting substance, the maximum dry specific weight of clayey sand soil decreases but when the percentage of water is zero with an increase the hydrocarbon material from the dirt filter residue and acidic sludge to the clayey sand soil to 4.2% and 5.1%, respectively, the maximum dry specific weight of the soil increased. The reason for this could be the ease of movement of the soil particles relative to each other after they were contaminated with acidic sludge, which reduced the void ratio of the soil. A further increase in the percentage of contaminants filling the pores between the soil grains caused a decrease in the specific dry weight of the soil.

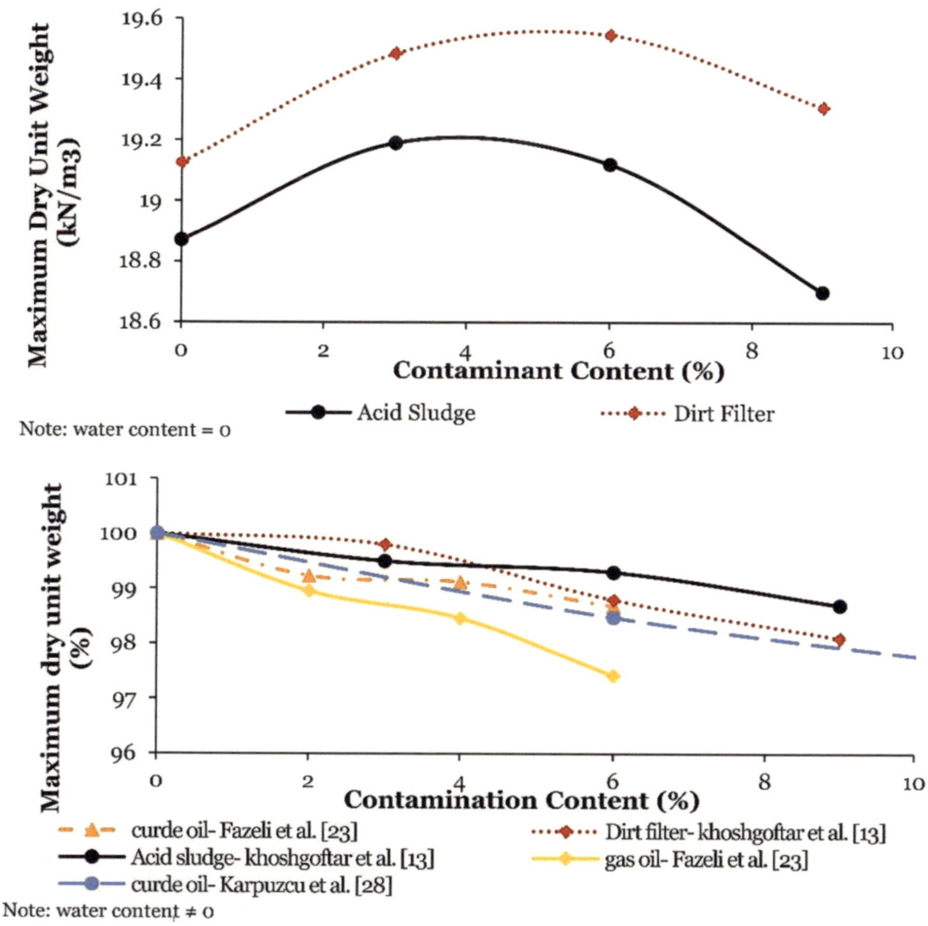

Figure 2.
Maximum specific dry weight of clayey sand soil vs. contamination content: (a) in current research; (b) in current and other research.

Table 2 show that with the contamination of the soil by hydrocarbon substances and its derivatives, the maximum dry weight of the soil and the optimum moisture content generally decreased.

Shin et al. [30] and Baiazidi et al. [19] studied sandy soil contaminated with gas oil and stated that, when gas oil surrounded the soil particles, the capillarity created between the contaminated particles and water caused the soil grains to become trapped. Because the water and pollutant both have low densities and are almost incompressible compared to soil, this caused a decrease in the maximum dry weight of the soil. The decrease in the optimum moisture percentage resulted from the space between the soil grains being filled with pollutant. The soil then reached the maximum dry weight at a lower optimum moisture percentage. This caused an improvement in the compaction parameters of the soil [13, 31].

In fine-grained soil, because petroleum compounds have a very low dielectric constant and exhibit non-polarity, they become separated from the surface of the clay

References	Soil type	Type of contamination	Content (%)	curing period	MDD	OMC
Khoshgoftar et al. [13]	SC, Iran	Acid sludge	0–3–6–9	One day	FID	—
Fazeli et al. [23]	SC, Iran	Gas oil	0–2–4–6	One month	D	D
		Crude oil			D	D
Safehian et al. [24]	CH, Iran	Gas oil	0–4–8–12–16–20	A week	D	FID
Khamehchiyan et al. [25]	SP, Iran	Crude oil	0–4–8–12–16	One month	D	D
	SM				D	D
	CL				D	D
Nasehi et al. [21]	SP, Iran	Gas oil	0–3–6–9	One month	D	D
	CL				D	D
	ML				D	D
Abdelhalim et al. [26]	SM, Malaysia	Engine oil	0–3–5–8–10–15–20	Three days	I	D
	SC				I	D
Yazdi and Sharifi [27]	ML, Iran	Gasoline	0–3–6–9–12	A week	D	D
				One month	D	D
Karpuzcu et al. [28]	SC, Turkey	Crude oil	0–6–12	two weeks	D	D
	CL				D	D
Hamidi and Karimi [20]	SC-SM, Iran	Crude oil	0–3–5–7	A week	D	D
Karkush and Jihad [29]	CH, Iraq	Kerosene	0–10–20–30–40–50–60	One month	I	D

Note: MDD, maximum dry density; OMC, optimum moisture content; FID, first increased and then decreased; I, increased; D, decreased.

Table 2.
Optimal moisture percentage and maximum specific dry weight of soil types contaminated with hydrocarbon and non-carb on materials.

grains and the double layer and move into the pore fluid, causing a gap between the clay particles and acting as a load insulator. This act is electrical and has a negative effect on the electrostatic attraction field of clay particles (fine-grained soil acts like coarse-grained soil) and contributes to a reduction in the amount of water required for compaction of samples [23].

The addition of a hydrocarbon to the soil will produce about half of the amount of soil compaction that the addition of water will because water molecules are bipolar and hydrocarbon molecules are non-polar and the clay particles do not tend to absorb them [24].

3.2 Effect of contamination on soil resistance parameters

3.2.1 Internal friction angle and soil cohesion (direct shear test)

As the soil particles were contaminated with acidic sludge, the contaminant settled between the soil particles and bind to non-polar acidic sludge compounds and active soil surface substances, including SiO2, by Van Der Waals interactions (**Figure 3**). An increase in the contaminant content increased the Van Der Waals force, leading to an increase in soil cohesion. Soil cohesion also increased with an increase in the dirt filter contaminant content owing to the inherent structure of the contaminants. The dirt filter contaminants behaved like fine-grained soil when added to granular soil, which resulted in an increase in soil cohesion. The increasing trend in soil cohesion is presented in **Table 3** with increases in the contaminant contents. Other researchers also have reported an increase in soil cohesion due to the high viscosity of hydrocarbon materials, but some researchers also have reported a decrease caused by contamination of the soil with hydrocarbons. Karkush and Kareem [29] reported a

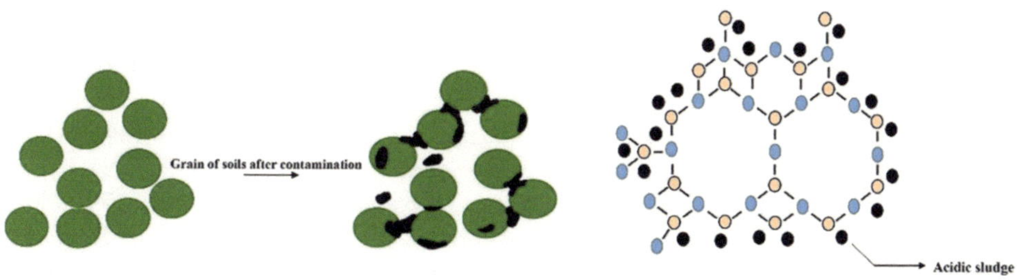

Figure 3.
Soil particles before and after contamination with acidic sludge.

Oil type	Vertical stress (kPa)	Contaminant content			
		0	3	6	9
Acid sludge	10–30–50	0.13	0.225	0.167	0.192
	50–100–150	0.06	0.207	0.229	0.258
Dirt filter	10–30–50	0.129	0.16	0.182	0.342
	50–100–150	0.06	0.393	0.364	0.5

Table 3.
Soil cohesion resulting from dirt filter residue and acidic sludge.

Oil type	Vertical stress (kPa)	Contaminant content			
		0	3	6	9
Acid sludge	10–30–50	49	41	39	38
	50–100–150	47	41	37	33
Dirt filter	10–30–50	49	48	44	43
	50–100–150	47	45	34	34

Table 4.
Soil internal friction angle resulting from dirt filter residue and acidic sludge.

decrease in the cohesion of clay soil with low plasticity properties when contaminated with 20% crude oil. This decrease could be due to a decrease in the dielectric constant and in the double layer thickness of the clay soil and change the structure of the soil to a scattered state (fine-grained soil will resemble granular soil).

Table 4 shows changes in the internal friction angle of the soil with the addition of both acidic sludge and dirt filter residue. As shown, the acidic sludge hydrocarbons moved between the soil particles and eliminated direct contact between them. As a consequence, the soil particles were able to slide relative to one another, leading to a reduction in the soil friction angle. Also, the addition of dirt filter residue caused the soil grains to become surrounded by a cover of filter particles, which reduced the roughness of the soil particles and the internal friction angle of the soil.

Table 5 show the results for other contamination soils. The changes in the internal friction angle and cohesion of the contaminated soil generally decreased, depending on the type of soil and the pollutant. For granular soil, lubrication between the grains which caused sliding of the grains and a decrease in the internal friction angle. Saberian and Khabiri [33] reported an 18% decrease in the internal friction angle with the addition of 8% of gas oil to poorly grained sandy soil.

Khoshgoftar et al. [13] reported that the addition of 9% acidic sludge and dirt filter residue from burnt oil treatment decreased the tightness of the connection between the soil particles. The roughness of the soil particles and the ease with which the soil grains slid relative to one another caused a 24% decrease in the internal friction angle at a compression level of 1.7 g/cm_3 under vertical stresses of 50, 100 and 150 kPa. In fine-grained soil, the internal friction angle generally increased because of flocculation of the clay soil with the increase of hydrocarbon pollutants [25]. Clay with a low plasticity property of 16% showed an 8° increase in the internal friction angle.

3.2.2 California bearing ratio (CBR)

One of the most important soil resistance tests for evaluation of contaminated soil for use in road construction is the CBR test. With an increase in the percentage of hydrocarbon pollutants in the soil, the CBR will generally decrease because of the effect of lubrication between the soil grains which allows them to slide relative to each other. **Table 6** shows some of the most important studies in this area. As can be seen, the soil resistance generally decreased in soil contaminated with hydrocarbons and the granular and clay soils behaved differently.

Figure 4 shows the CBR values for sandy and clay soils contaminated with burnt engine oil, which revealed that the CBR for all types of sandy soil initially increased and then decreased. In clay soil, with a high specific surface area for the clay particles,

References	Curing period	Vertical stress (kPa)	Cont (%)	Type of contamination	Soil type	φ (°)	C (kPa)
Khoshgoftar et al. [13]	One day	50–100–150	0	Acid sludge	SC	42	11
			3			35	13.5
			6			34	21
			9			34	26.3
			0	Dirt filter		42	11
			3			41	26.5
			6			40	26.7
			9			34	16.9
Shirdel and Makarchian [32]	A week	50–75–100	0	Crude oil	CL	27	24.2
			10			27	28.3
			20			31	27.9
			30			24	27
			0	Gas oil		27	24.2
			10			29	23.4
			20			35	22.6
			30			*	*
Abdelhalim et al. [26]	Three days	14.3–24.1–43.7	0	Motor oil	SM	40	17.5
			3			37	17.9
			5			33	18.6
			8			32	18.6
			10			29	19.1
			0		SC	35	35
			3			30	35.7
			5			28	33.9
			8			27	31.3
			10			25	26.4
			9			32	6.8
			12			31	7.1
			15			30	7.3

Table 5.
Cohesion and internal friction angles of soil contaminated with hydrocarbons by maximum specific dry weight density.

the amount of pollutant surrounding the particles per unit volume was higher and caused more slippage of particles and, subsequently, a greater decrease in the CBR compared to granular soil.

3.3 Effect of oil contamination on Atterberg limits

When the soil is contaminated with the hydrocarbons found in oil and its derivatives, changes in the repulsive force, cation exchange capacity and viscosity cause

References	Soil type	Type of contamination	Content (%)	curing period	CBR
Abdelhalim et al. [26]	SM	Motor oil	0–3–5–8–10	Three days	D
	SC				FDI
Al-Sanad et al. [34]	SP	Crude oil	0–2–4–6		FID
Rasheed et al. [35]	SPSM	Gas oil	0–3–5–7.5	One day	D
		Kerosene			
Choura et al. [36]	SP	Crude oil	0–5–10–15		I
Alhassan and Fagge [37]	SP	Crude oil	0–2–4–6		D
		Used engine oil			D
	CL	Crude oil			D
		Used engine oil			D
Karpuzcu et al. [28]	CL	Crude oil	0–6–12	Two weeks	D
	SC				I

Note: D, decreased; I, increased; FDI, first decreased and then increased; FID, first increased and then decreased.

Table 6.
CBR of soil contaminated with hydrocarbons.

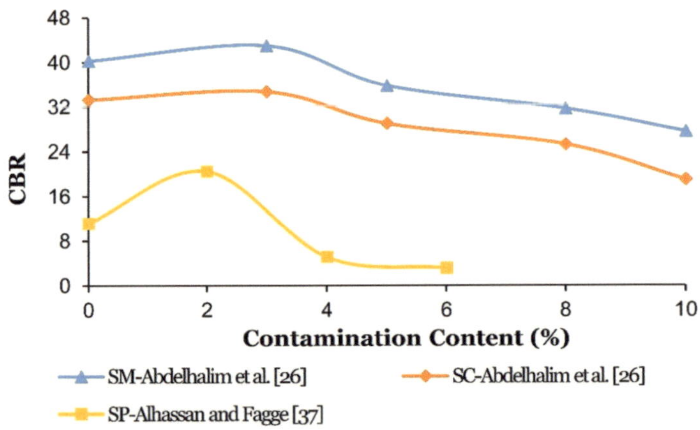

Figure 4.
CBR vs. increase in burnt oil residue for all soil types.

changes in the plastic limit, liquid limit and plasticity index of the soil. **Table 7** show the Atterberg limits by the type of soil contaminated based on the type of pollutant, duration of pollution and type of soil. In general, with an increase in the soil contamination, leading to an increase in cohesion between the soil grains, the Atterberg limits increased. With an increase in pollution, the Atterberg limits decreased slightly and then remained the same. With an increase in the length of the contamination period, the Atterberg limits decreased.

As can be seen in **Figure 5**, with the passage of time (about 1 month) after gasoline pollution, the liquid limit of the soil decreased compared to a shorter period (one week) [27].

References	Soil type	Type of contamination	Content (%)	Curing period	PL	LL	PI
Fazeli et al. [23]	SC	Gas oil	0–2–4		D	D	D
	Iran	Crude oil			D	D	D
Roshanqhiyas and Bagheripour [38]	CL Iran	Crude oil	0–2–4–6–8–10	One month	D	D	D
Abdelhalim et al. [26]	SC Malaysia	Motor oil	0–3–5–8–10	Three days	D	D	D
	SM Malaysia				D	D	D
Kermani and Ebadi [7]	CL Iran	Crude oil	0–4–8–12	A week	I	I	D
Karkush and Kareem [29]	CL Iraq	Fuel oil	0–10–20	Four days	I	D	D
Hamidi and Karimi [20]	SC-SM Iran	Crude oil	0–3–5–7	A week	I	I	I
Eissa [39]	CH Egypt	Gasoline	0–2–4–8–10–16	Ten days	D	D	D
Karkush and Jihad [40]	CH Iraq	Kerosene	0–10–20–30–40–50–60	One month	I	I	I

Note: PL, plastic limit; LL, liquid limit; PI, plastic index; D, decreased; I, increased.

Table 7.
Atterberg limits of soil types contaminated with hydrocarbon and non-carbon materials.

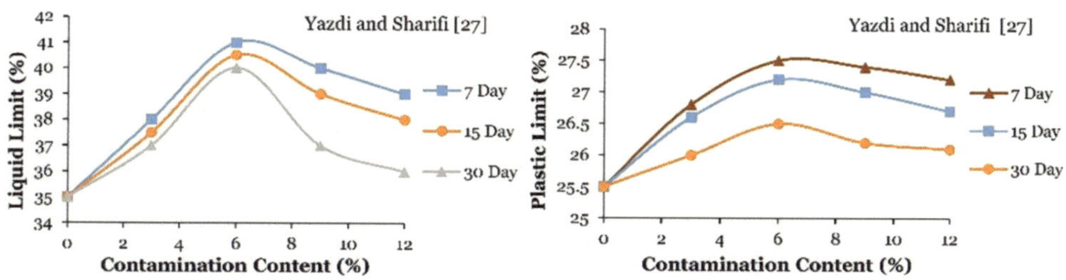

Figure 5.
Liquid and plastic limits of ML soil contaminated with gasoline.

3.4 Effect of oil contamination on soil permeability

Table 8 show that, with the contamination of all types of soil with hydrocarbons, the permeability generally decreases. Iloeje and Aniago [42] recorded a decrease 27% with 8% crude oil contamination of clay soil. The reason for the decrease in permeability can be due to the filling of the space between the soil grains by the hydrocarbon materials, which are of high viscosity and make movement through the space between the soil grains more difficult. Baiazidi et al. [19] and Gillot [44] reported that, because of the hydrophobic nature of hydrocarbons, they form an impermeable curtain in front of the water and some of the water's energy is used to tear this

References	Soil type	Type of contamination	Content (%)	Results	Mechanism
Baiazidi et al. [19]	GM Iran	Diesel	0–3–6–8–11	D	Due to the hydrophobicity of hydrocarbon materials and the swelling of clay particles
	SM Iran			D	
Oyediran and Enya [41]	SW Nigeria	Crude oil	0–10	D	Due to the high viscosity of crude oil and the filling of the empty space between the grains
Hamidi and Karimi [20]	SC-SM Iran	Crude oil	0–3–5–7	I	With the increase in pollution, due to the decrease in the dielectric constant and the thickness of the double layer of the coagulated soil grains, the space between the grains increases.
Iloeje and Aniago [42]	CL Nigeria	Crude oil	0–2–4–6–8	D	Filling the empty space between the grands
Jedari and Farahani [43]	CL Iran	Kerosene	0–3–6–9	D	Due to the high viscosity of hydrocarbon materials, water flow becomes difficult
		Gas oil		D	
	CH Iran	Kerosene		D	
		Gas oil		D	

Note: D, decreased; I, increased.

Table 8.
Permeability of soil types contaminated with hydrocarbon substances.

hydrocarbon layer. They also stated that, in fine-grained soil, when water-soluble materials enter the water-rich soil particles, water and other cations that surround the water-rich layer (double layer of clay minerals) may be absorbed by the components. The substitutes for the organic molecules can enter the interlayer spaces and cause swelling of the clay layer.

Some researchers have reported that, with soil pollution by hydrocarbons, the dielectric constant and the thickness of the double layer will decrease, which will cause flocculation of the soil particles and increase the space and permeability between the grains [20, 45]. As this pollutant has a high pH, soil particles could dissolve in it, which would increase the space between the grains and the soil permeability.

4. Conclusion

The effect of contamination on the mechanical properties of soil were reviewed to evaluate their suitability for engineering applications and the following conclusions can be drawn:

- The rate of change of some of the geotechnical parameters of contaminated soil decreased with the passage of a long time (one month compared to one week), but the rate of change of other parameters increased.

- Contamination with 4.2% acidic sludge and 5.1% dirt filter residue produced the maximum specific dry weight of the soil. Further increases in the amount of pollutant caused filling of the pores between the soil grains and caused the rate of change to decrease.

- The low dielectric constant and non-polarity of hydrocarbon pollutants reduced the double layer thickness in clay, which caused it to flocculate and form a granular soil structure.

- With the contamination of clay soil, the changes of Atterberg limits were generally increased.

- In granular soil, the pollutant surrounded the soil grains and, because of their high viscosity, caused lubrication of the soil grains, which reduced the internal friction angle and increased the compaction parameters of the soil.

- The shear strength of soil contaminated with hydrocarbons showed no definite and constant trend of change. Acidic sludge as a pollutant, despite the increase in soil adhesion, caused a decrease in the internal friction angle of the soil and increased its shear resistance. Dirt filter residue as a pollutant increased the shear resistance of the soil because of the large increase in adhesion, despite the decrease in the internal friction angle.

- The high viscosity of hydrocarbons in the soil generally increased soil cohesion.

- Contamination by hydrocarbon substances caused filling of the empty spaces between the soil grains. This, along with the high viscosity of the petroleum substances caused a general decrease in permeability.

Author details

Mahdi Khodaparast[1*] and Amir Khoshgoftar[2]

1 Faculty of Engineering, Department of Civil Engineering, University of Qom, Iran

2 Faculty of Engineering, Geotechnical Engineering, Department of Civil Engineering, University of Qom, Iran

*Address all correspondence to: khodaparast@qom.ac.ir

References

[1] Tumanyan A, Tyutyuma N, Bondarenko A, Shcherbakova N. Influence of oil pollution on various types of soil. Chemistry and Technology of Fuels and Oils. 2017;**53**(3):369-376. DOI: 10.1007/s10553-017-0813-7

[2] Errington I, King CK, Houlahan S, George SC, Michie A, Hose GC. The influence of vegetation and soil properties on springtail communities in a diesel-contaminated soil. Science of the Total Environment . 2018;**619-620**:1098-1104. DOI: 10.1016/j.scitotenv.2017.11.186

[3] Haghsheno H, Arabani M. Geotechnical properties of oil-polluted soil: A review. Environmental Science and Pollution Research International. 2022;**29**(22):32670-32701. DOI: 10.1007/s11356-022-19418-1

[4] Prasanna G, Manoharan S. A review on effect of crude oil on the geotechnical properties of soil. International Journal of Research in Engineering and Technology. 2016;**3**:1234-1236

[5] Tajik M. Assessment of geoenvironmental effect of petroleum pollution on coastal sediments of bushehr province-iran [thesis]. Tehran, Iran: Tarbiat Modares University; 2004 (in Persian)

[6] Tuncan A, Pamukcu S. Predicted mechanism of crude oil and marine clay interactions. In: Usmen A, editor. Environ Geotech. Rotterdam: Balkema; 1992. pp. 109-121

[7] Kermani M, Ebadi T. The effect of oil contamination on the geotechnical properties of fine-grained soils. Soil and Sediment Contamination: An International Journal. 2012;**21**(5):655-671. DOI: 10.1080/15320383.2012.672486

[8] Riser E. Remediation of Petroleum Contaminated Soils: Biological, Physical, and Chemical Processes. New York, United States of America: CRC Press; 1998

[9] Al-Rawas AA, Hago AW, Al-Sarmi H. Effect of lime, cement and Sarooj (artificial pozzolan) on the swelling potential of an expansive soil from Oman. Building and Environment. 2005;**40**:681-687. DOI: 10.1016/j.buildenv.2004.08.028

[10] Virkutyte J, Sillanpää M, Latostenmaa P. Electrokinetic soil remediation – Critical overview. Science of The Total Environment. 2002;**289**:97-121. DOI: 10.1016/S0048-9697(01)01027-0

[11] Jukic A. Petroleum Refining and Petrochemical Processes Crude Oil: Composition, Classification. Croatia: Faculty of Chemical Engineering and Technology University of Zagreb; 2013. pp. 37-46

[12] Cyrus S, Kumar T, Abraham B, Sridharan A, Jose B. Effect of industrial wastes on the physical and engineering properties of soils. Proceedings of Indian Geotechnical Conference GEOtrendz. 2010;**1**:16-18

[13] Khoshgoftar A, Khodaparast M, Sedighi M. Effect of residues from a burnt oil refinery on the compaction parameters and strength of clayey sand. Bulletin of Engineering Geology and the Environment. 2021;**80**(8):6331-6341. DOI: 10.1007/s10064-021-02320-4

[14] Fine P, Graber E, Yaron B. Soil interactions with petroleum hydrocarbons: Abiotic processes.

Soil Technology. 1997;**10**(2):133-153. DOI: 10.1016/S0933-3630(96)00088-8

[15] Nudelman N, Rios I, Katusich O. Fate of the oil residuals in Patagonian soils effects of the environmental exposure time. Journal of Environmental Risk Assessment and Remediation. 2002;**3**:1-8

[16] Singh S, Srivastava R, John S. Studies on soil contamination due to used motor oil and its remediation. Canadian Geotechnical Journal. 2009;**46**(9):1077-1083. DOI: 10.1139/T09-047

[17] Mohammadi M, Khodaparast M, Rajabi AM. Effect of nano calcium carbonate (nano $CaCO_3$) on the strength and consolidation properties of clayey sand soil. Road Materials and Pavement Design. 2021;**23**(10):2394-2415. DOI: 10.1080/14680629.2021.1976255

[18] Mohammadi M, Rajabi AM, Khodaparast M. Experimental and numerical evaluation of the effect of nano calcium carbonate on geotechnical properties of clayey sand soil. KSCE Journal of Civil Engineering. 2022;**26**:35-46. DOI: 10.1007/s12205-021-1914-8

[19] Baiazidi O, Mohammadi SD, Tajik M. Investigation of gasoil contamination effect on the engineering properties of surface soils around the hamedan oil storage [thesis]. Hamedan, Iran: University of BUAli Sina Faculty of Sciences Department of Geology; 2013 (in Persian)

[20] Hamidi A, Karimi AH. Effect of phytoremediation on compression characteristics of silty clayey sand contaminated with crude oil. International Journal of Civil Engineering. 2021;**19**(8):995-973. DOI: 10.1007/s40999-021-00609-9

[21] Nasehi SA, Uromeihy A, Nikudel MR, Morsali A. Influence of gas oil contamination on geotechnical properties of Fine and coarse-grained soils. Geotechnical and Geological Engineering. 2016;**34**(1):333-345. DOI: 10.1007/s10706-015-9948-7

[22] ASTM D698-12E2. Standard Test Methods for Laboratory Compaction Characteristics of Soil Using Standard Effort (12 400 Ft-Lbf/ft3 (600 kN-M/m3)). West Conshohocken, PA: ASTM (American Society for Testing and Materials); 2007. DOI: 10.1520/D0698-12E02

[23] Fazeli G, Farrokhi F, Firoozfar A. Determination and evaluation of bearing capacity of clay soil contaminated with oil pollutants and studying the geotechnical characteristics of the substrate [thesis]. Zanjan, Iran: University of Tabriz Faculty of Civil Engineering; 2019 (in Persian)

[24] Safehian H, Rajabi AM, Ghasemzadeh H. Effect of diesel-contamination on geotechnical properties of illite soil. Engineering Geology. 2018;**241**:55-63. DOI: 10.1016/j.enggeo.2018.04.020

[25] Khamehchiyan M, Charkhabi AH, Tajik M. Effects of crude oil contamination on geotechnical properties of clayey and sandy soils. Engineering Geology. 2007;**89**(3-4):220-229. DOI: 10.1016/j.enggeo.2006.10.009

[26] Abdelhalim RA, Ramli H, Selamat MR. Investigating the impacts of oil contamination on geotechnical properties of laterite soils. Innovative Infrastructure. Solutions. 2022;**7**(321):1-14. DOI: 10.1007/s41062-022-00901-0

[27] Yazdi A, Sharifi E. Effects of contamination with gasoline on engineering properties of fine-grained silty soils with an emphasis on the duration of exposure. SN Applied

Sciences. 2021;**3**(7):1-31. DOI: 10.1007/s42452-021-04637-x

[28] Karpuzcu M, Arbili M, Ghaffoori F, Cabalar A. An assessment of the effects of light crude oil on the geotechnical properties of Asoil. International Journal of Advances in Mechanical and Civil Engineering. 2018;**5**:1-4

[29] Karkush MO, Kareem ZA. Investigation of the impacts of fuel oil on the geotechnical properties of cohesive soil. Engineering Journal. 2017;**21**(4):127-137. DOI: 10.4186/ej.2017.21.4.127

[30] Shin EC, Lee IB, Das BM. Bearing capacity of a model seal footing on crude oil contaminated sand. Geotechnical and Geological Engineering. 1995;**17**(9):1-12

[31] Rajaee F, Lashkaripour GR, Nikudel MR, Ghfoori M. Investigation of engean-oil contamination effect ongeothechnical properties of soil [thesis]. Mashhad, Iran: Ferdowsi University; 2012 (in Persian)

[32] Shirdel M, Makarchian M. Investigation of the impact of considerable oil contamination on mechanical properties of fine-grained soils. Marine Engineering. 2020;**16**(3):97-106. DOI: 10.29252/marineeng.16.32.97

[33] Saberian M, Khabiri MM. Effect of oil pollution on function of sandy soils in protected deserts and investigation of their improvement guidelines (case study: Kalmand area, Iran). Environmental Geochemistry and Health. 2016;**40**(1):243-254. DOI: 10.1007/s10653-016-9897-y

[34] Al-Sanad HA, Eid WK, Ismael NF. Geotechnical properties of oil-contaminated Kuwaiti sand. Journal of Geotechnical Engineering. 1995;**121**(5):407-412. DOI: 10.1061/(ASCE)0733-9410(1995)121:5(407)

[35] Rasheed Z, Ahmed F, Jassim H. Effect of crude oil products on the geotechnical properties of soil. WIT Transactions on Ecology and the Environment. 2014;**186**:353-361. DOI: 10.2495/ESUS140301

[36] Choura M, Salhi S, Cherif F. Mechanical behaviour study of soil polluted by crude oil: Case of Sidi El Itayem oilfield, Sfax. Tunisia. Environmental Earth Sciences. 2009;**59**(3):573-580. DOI: 10.1007/s12665-009-0055-z

[37] Alhassan HM, Fagge SA. Effects of crude oil, low point pour fuel oil and vacuum gas oil contamination on the geotechnical properties sand, clay and laterite soils. International Journal of Engineering Research and Applications. 2013;**3**(1):1947-1954

[38] Roshanqhiyas M, Bagheripour M. Effect of crude oil contamination on the geotechnical properties of kaolinite clay soil in the road. Transportation Infrastructure Engineering. 2019;**5**:101-112. DOI: 10.22075/JTIE.2019.17368.1371

[39] Eissa AM. Effects of gasoline-contamination on geotechnical properties of clayey soil. African Young Geotechnical Engineers Conference. 2016;**16**(12):1-11

[40] Karkush MO, Jihad AG. Studying the geotechnical properties of clayey soil contaminated by kerosene. Key Engineering Materials. 2020;**857**:383-393. DOI: 10.4028/www.scientific.net/KEM.857.383

[41] Oyediran I, Enya N. Crude oil effects on some engineering properties of sandy alluvial soil. International Journal of Mining and Geological Engineering. 2022;**56**:7-10. DOI: 10.22059/IJMGE.2020.283051.594815

[42] Iloeje A, Aniago V. Effect of crude oil on permeability properties of the soil. International Journal of Trend in Scientific Research and Development. 2016;**1**(1):39-43. DOI: 10.31142/ijtsrd5724

[43] Jedari C, Farahani M. Permeability and compression characteristics of clay contaminated with kerosene and gasoil. International Journal of Architecture, Urbanism and Civil Engineering. 2018;**1**(3):1-10. DOI: 10.33544/mjauce.v1i3.72

[44] Gillott JE. Clay in Engineering Geology. 2nd ed. The University of Cagary, Canada: Elsevier Science Limited; 1987

[45] Ostovar M, Ghiassi R, Mehdizadeh MJ, Shariatmadari N. Effects of crude oil on geotechnical specification of sandy soils. Soil and Sediment Contamination: An International Journal. 2021;**30**(1):58-73. DOI: 10.1080/15320383.2020.1792410

Chapter 2

A Review of the Effect of Hydrocarbon Contamination on Soil Resistance Parameters

Mahdi Khodaparast and Amir Khoshgoftar

Abstract

Oil contamination and derivatives are some of the types of pollution that can lead to irreparable damage to the environment. Oil pollution affects the mechanical, chemical, and dynamic properties of the soil. Petroleum contaminants alter the structure of cohesive and granular soil. The double-layer thickness of clay will change, causing the structure of the clay soil to become similar to granular soil. The results of adding oil contaminants to the soil showed decreases in the internal friction angle and increase in the cohesion of the soil. The shear strength of the contaminated soil showed no definite or constant trend of change. When contaminated with acidic sludge (burned-oil waste derivatives), despite an increase in the cohesion of the soil, a decrease in the internal friction angle caused a decrease in the shear strength. The dilatancy angle decreased as the dirt filter contaminant content increased. In more compact samples, the soil dilatancy angle remained constant after the addition of 3% dirt filter contaminants at high and low vertical stresses and did not affect the declining trend for shear strength.

Keywords: hydrocarbon contamination, soil resistance, geotechnical properties, soil dilatancy angle, direct shear test

1. Introduction

About 83.6 million barrels of crude oil are produced in the world daily and a significant portion of it penetrates to the soil due to different reasons. This infiltration can occur by leaking of the transmission pipes or overturning of the crude oil transmission tankers. The soil of refineries is contaminated with oil materials for various reasons [1]. Biological, Washing, physical, and chemical methods have been introduced to clean contaminated soil, but they are costly and have limitations [2]. In some cases, due to the large volume of the operation of moving contaminated soil and the difficulty and uneconomicality of cleaning contaminated soil, the same contaminated soil can be used, and the shear strength parameters were considered as a criterion for subsequent designs [3, 4].

Soil contamination with hydrocarbon substances can change the physical, chemical, and resistance parameters of the soil. So far, there have been many studies on the impact of various types of oil pollution on fine- and coarse-grained soils, and sometimes different results have been obtained. These differences are mainly due to the different reactions that different types of soil show against different types of pollution [5]. The poorly graded sandy soil contaminated with crude oil was examined, and it was found that, as the percentage of oil contaminants increased, the shear strength decreased. Crude oil and its derivatives are mainly composed of hydrocarbons, nitrogen compounds, sulfur, organic metal compounds, and mineral salts [6]. The duration of contamination, the type of pollutant, and the type of soil will affect the geotechnical properties of the polluted soil [7]. Any change in the shear resistance and other engineering characteristics of contaminated soil layers can cause a decrease in the stability of slopes, decrease the bearing capacity, and increase the total and relative settlement of structural foundations [8].

Nasehi et al. [9], by performing a direct shear test on the ML, CL, and SP soil contaminated with gas oil, showed a decrease in the optimum moisture content, soil friction angle, and relative soil density and an increase in soil cohesion with an increase in the contaminant content. Also, Saberian and Khabiri [10] stated that with sand pollution to 8% of gas oil, the optimum moisture content decreased by 4 percent and increased the maximum dry unit weight from 18.25 to 19.27 kN/m^3.

Karkush and Kareem [11] stated that by performing a direct shear test on clayey soil contaminated with 20% crude oil, the internal friction angle of the soil decreases by 12 degrees. They also discovered a 28.5% reduction in undrained soil cohesion. Kermani and Ebadi [12] tested clay soil contaminated with oil and reported that increased oil contamination decreased soil cohesion.

The soil investigated in this chapter has been exposed to pollutants and waste from a petroleum factory. This included the waste from burned oil from a refinery. Different studies have been conducted on the impact of hydrocarbons and petroleum pollutants, including crude oil, gasoline, diesel, motor oil, and burned motor oil in fine and coarse soils, but little study has been done on soil contamination by waste from burned oil from refineries. Burned oil refining factories bury this substance, which causes soil pollution. A literature review showed that the contamination of different types of soil by hydrocarbons would cause different changes in the soil geotechnical parameters. The results of these studies should be compared and analyzed; therefore, a complete and accurate study of the mechanisms of hydrocarbon contamination and its derivatives on the characteristics of the soil is required. This chapter was conducted to investigate differences in the process of change in the geotechnical parameters of all types of soil contaminated with hydrocarbons, especially acidic sludge and dirt filter residue, both waste products from a burned oil refinery.

2. Data and method

In previous studies, different types of soil with different weight percentages were mixed with crude oil and its derivatives and then tested to determine the impact of the pollutant on the resistance soil, compaction, soil settlement, etc. of the soil. In this chapter, the results of changes in the parameters of the contaminated soil were extracted using the data available from other research with the help of appropriate software (Get Data Graph Digitizer) and evaluated and analyzed. And secondly, in order to identify the contamination effect of materials obtained from the burning oil treatment on the compaction parameters of the soil, a modified proctor test was performed according to standard ASTM D698 [13] as well as a 100 × 100 mm sample box direct shear test to check the trend of soil resistance parameters according to standard ASTM D3080 [14] with contamination percentages of 0, 3, 6, and 9.

The choice of these pollutant values is to investigate the effect of low pollution on the geotechnical properties of the soil because when the contamination is high, it is taken into consideration and cleaned, while it may be ignored in low pollution percentages. Shroff [15] also showed that the maximum percentage of oil contamination in soils is about 10%.

The shear test was performed with a vertical stress of 10, 30, and 50 kPa with dry unit weight 1.7 and 1.9 g/cm^3 to investigate the effect of the depth of contaminated soil according to **Table 1**.

2.1 Soil properties and study area

The studied soil was collected from the landfill site of the waste from the burned oil refining factories located in the industrial city of Salafchegan. As shown in **Figure 1**, the city of Salafchegan is located 35 kilometers west of Qom. This city is located on the Qom- Arak road. The geological setting of the Saveh-qom basin is also presented in **Figure 2** [17]. The soil gradation (SC) test was performed according to the standard of ASTM D422 [18], and the characteristics of the studied soil are shown in **Table 2** (**Figure 3**).

2.2 Contaminant properties

In this chapter, firstly, different types of soil with different weight percentages were mixed with crude oil and its derivatives. Secondly, the waste material from the burned oil refining factories (acidic sludge and dirt filter) was used as a pollutant. The primary components of these two contaminants, as determined from x-ray fluorescence tests, were zinc, ferrum, plumbum, cuprum, phosphorus, dust, glycols, carbon, phenols, polymers, asphaltene, polymethacrylates, sulfonates, amines and phosphonates. The results of XRF testing are shown in **Table 3**, respectively.

Number of tests	Type/Value	Parameter
2	Acidic sludge	Contaminant type
	Dirt filter	
2	17	Dry unit weight (kN/m^3)
	19	
2	0.1–0.3–0.5	Vertical load steps (kg/cm^3)
	0.5–1–1.5	
4	0	Contamination content
	3	
	6	
	9	

Table 1.
Test conditions.

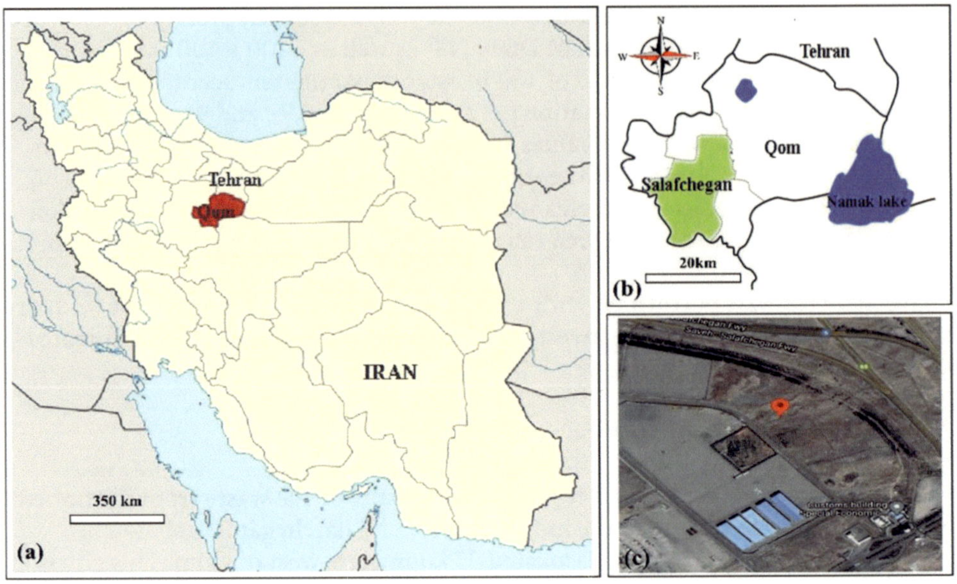

Figure 1.
(a) Position of the Qom city, (b) Position of the Salafchegan city, and (c) Satellite image of Salafchegan industrial region.

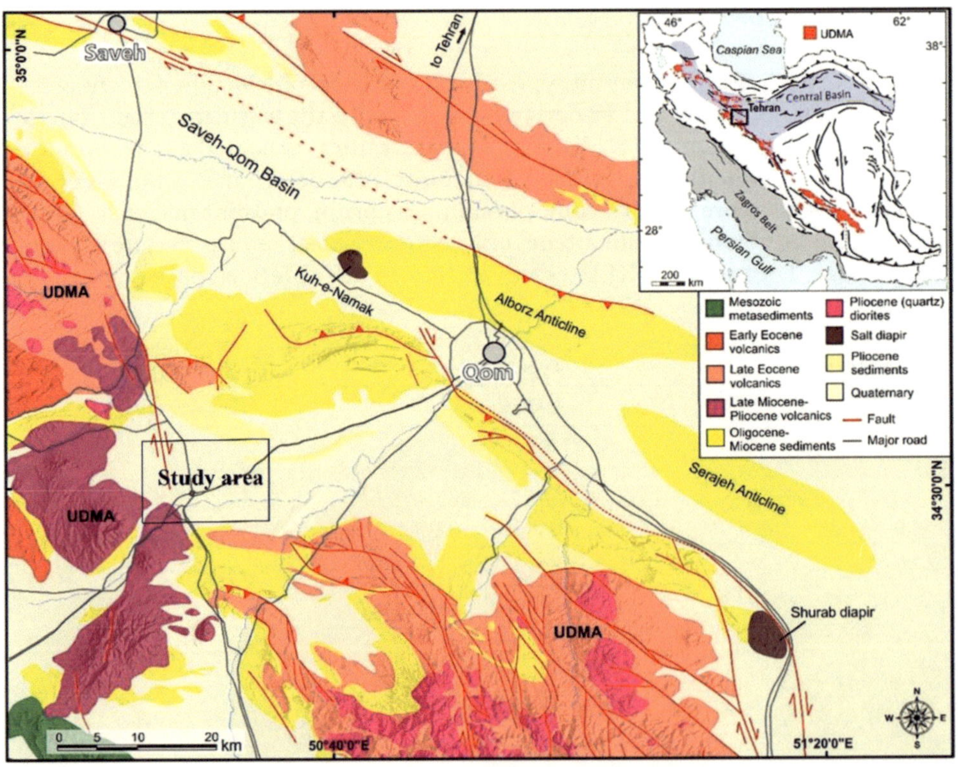

Figure 2.
Regional geology of a sector of Central Iran comprising the Salafchegan city [16].

NMC ω (%)	OMC ω	$\gamma_{d\,min}$ (kN/m^3)	$\gamma_{d\,max}$ (kN/m^3)	PL (%)	LL (%)	C_c	C_u	Parameter
3.51	8.9	14.72	20.48	15	30	4.52	331.92	Value

Table 2.
Soil properties.

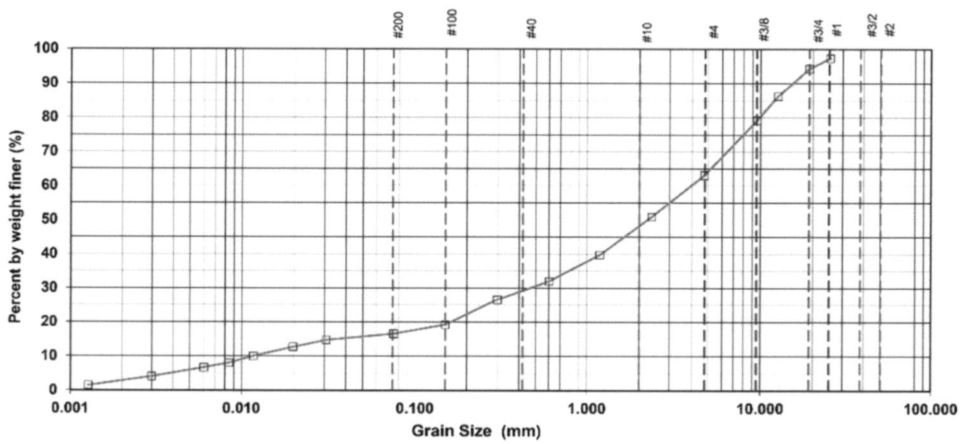

Figure 3.
Grain-size distribution curve of soil.

3. The effect of contamination on soil geotechnical parameters

The effect of pollution on granular soils is physical, and on cohesive soils, it is a change in soil texture. Generally, cohesive soil is more affected by oil contamination. Past research shows that soil compaction and resistance parameters decrease when hydrocarbon substances contaminate soil.

3.1 Compaction parameters

The modified compaction test was carried out on contaminated soil with zero moisture percentage in order to identify the effect of the contamination of materials resulting from the treatment of burned oil on the compaction parameters of the soil. **Figure 4** shows that with the increase of acid sludge contaminations up to 4.2% due to the reduction of void soil and easier sliding of grains soil, the maximum dry soil unit weight increases. But with increasing contamination due to the separation of grains of soil from each other, the maximum dry soil unit weight decreases.

With 5.1% contamination of the dirt filter, first, the specific weight of the soil increased, then due to the filling of the spaces between the soil grains, the specific weight decreased.

The results of other researchers are shown in **Table 4**, as it is clear that the specific dry weight of the soil decreased with the increase of contamination.

Figures 5 and **6** show that the addition of 12% crude oil to the soil caused a 2.5% decrease in the maximum dry weight and a 50% decrease in the optimum moisture content [25]. Karkush and Jihad [27] and Ghadyani et al. [24] stated that soil

	L.O.I	**Na2O**	**MgO**	**Al_2O_3**	**SiO2**	**P2O5**	**SO3**	**Cl**	**K2O**	**CaO**	**Cr**	**TiO2**
Acidic Sludge	86.87	0.425	0.09	0.079	0.322	1.207	6.683	0.008	0.099	2.941	0.01	
Dirt Filter	37.25	0.073	1.44	6.67	47.734	0.04	2.418	0.073	0.358	3.041		0.067
	Mn	*Fe2O3*	*Ni*	*Cu*	*Zn*	*Rb*	*Sr*	*Zr*	*Ba*	*Mo*	*Pb*	
Acidic Sludge	0.007	0.428	00.04	0.018	0.784		0.002	0.001		0.019	0.004	
Dirt Filter	0.028	0.736	0.004		0.007	0.002	0.016	0.005	0.04			

Table 3.
X-ray fluorescence test results on contaminants.

DOI: http://dx.doi.org/10.5772/intechopen.1006747

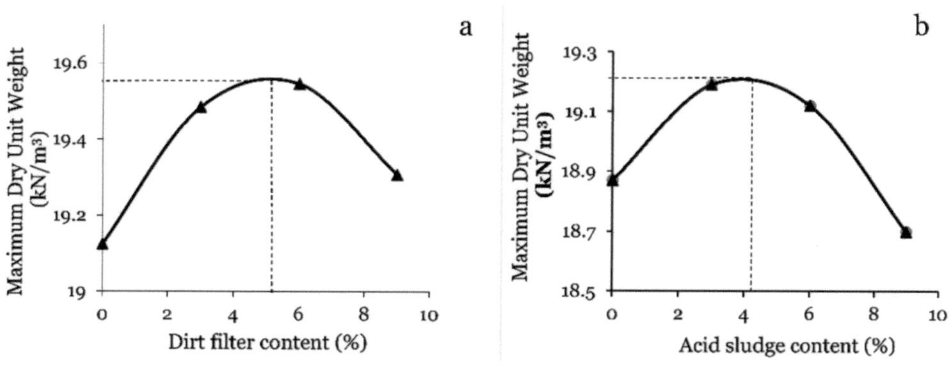

Figure 4.
Maximum dry unit weight of soil by percentage of (a) dirt filter, and (b) acidic sludge contaminants.

References	Soil type	Type of contamination	Content (%)	Curing period	MDD	OMC
Sarmadi et al. [19]	SP Iran	Kerosene	0–3–6–9–12–15	Two weeks	D	D
Fazeli et al. [20]	SC, Iran	Gas oil	0–2–4–6	One month	D	D
		Crude oil			D	D
George et al. [21]	SM India	Gas oil	0–4–8–12	A week	FID	
Khamehchiyan et al. [22]	SP, Iran	Crude oil	0–4–8–12–16	One month	D	D
	SM				D	D
	CL				D	D
Nasehi et al. [9]	SP, Iran	Gas oil	0–3–6–9	One month	D	D
	CL				D	D
	ML				D	D
Roshanqhiyas and Bagheripour [23]	CL Iran	Crude oil	0–2–4–6–8–10	One month	I	D
Ghadyani et al. [24]	CH Iran	Kerosene Gas oil	0–3–6–9	One month	I	D
					I	D
Karpuzcu et al. [25]	SC, Turkey	Crude oil	0–6–12	Two weeks	D	D
	CL				D	D
Choura et al. [26]	SP	Crude oil	0–5–10		D	D
Al-Sanad et al. [5]	SP Kuwait	Crude oil	0–2–4–6		I	D

Note: MDD: Maximum dry density, OMC: Optimum moisture content, FID: First increased and then decreased, I: Increased, D: Decreased.

Table 4.
Optimal moisture percentage and maximum specific dry weight of soil types contaminated with hydrocarbon and non-carbon materials.

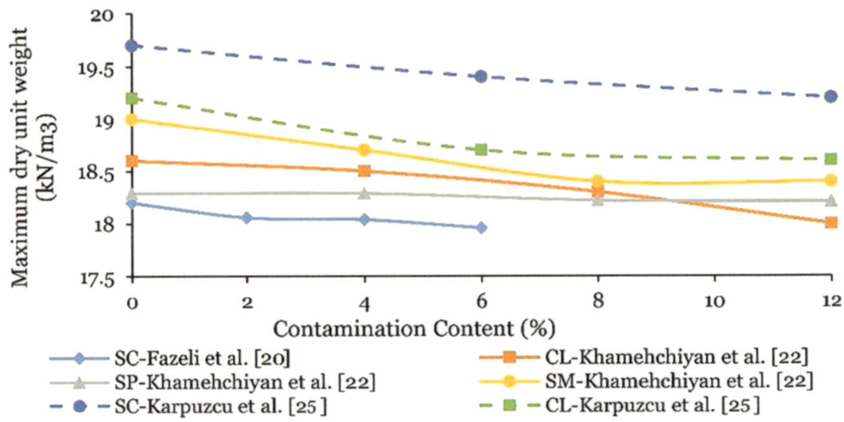

Figure 5.
Maximum specific dry weight of soil contaminated with crude oil.

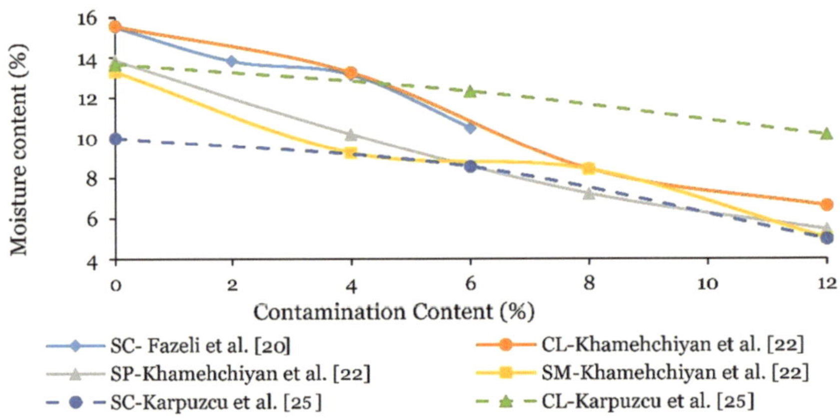

Figure 6.
Optimum moisture percentage of soil contaminated with crude oil.

contaminated by hydrocarbon substances would reach the maximum dry weight of the soil at a lower optimum moisture percentage.

3.2 Soil resistance parameters

3.2.1 The effect of contamination on the internal friction angle of the soil

Mohr-Columb rupture cover is generally curved, but it is linear in low vertical stresses, and the results presented by the researchers are also assumed to be linear. Usually, with the addition of hydrocarbon materials to the soil, these materials are placed between the soil grains and cause a decrease in the firm connection between the soil particles and the easier sliding of the soil grains on each other, and finally, the internal friction angle of the soil is reduced.

As it is clear in **Table 5**, with the increase of soil contamination, the friction angle of the soil decreases.

As shown in **Figure 7**, with an increase in vertical stress, the process of reducing the internal friction angle of the soil has increased. As the soil particles became

Dry unit weight (kN/m^3)	Vertical stress (kg/cm^2)	Contaminant content			
		0	3	6	9
(a)					
17	0.1–0.3–0.5	44	43	42	38
	0.5–1–1.5	42	41	40	34
19	0.1–0.3–0.5	49	48	44	43
	0.5–1–1.5	47	35	34	34
(b)					
17	0.1–0.3–0.5	44	42	41	41
	0.5–1–1.5	42	35	34	34
19	0.1–0.3–0.5	49	41	39	38
	0.5–1–1.5	47	41	37	33

Table 5.
Soil friction angle according to percentage of (a) dirt filter, and (b) acidic sludge contaminants.

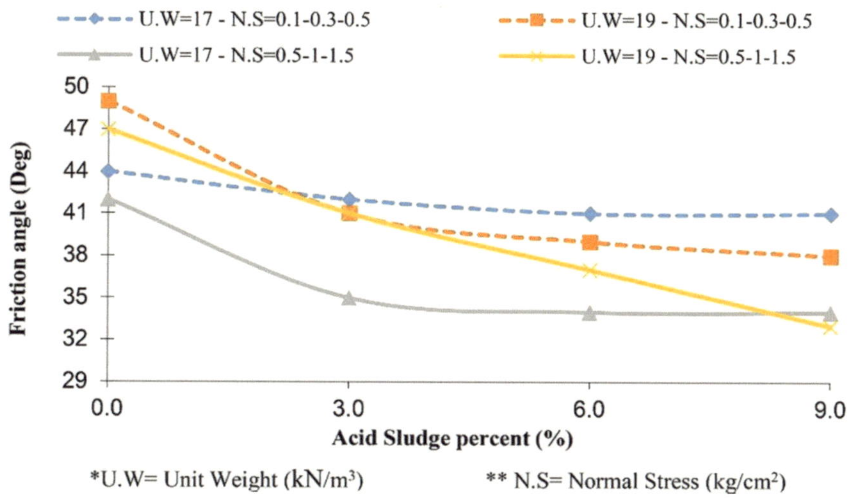

Figure 7.
Variation of internal friction angle of soil contaminated with different percentages of acidic sludge.

more compacted, acidic sludge had the function of a lubricant that reduced friction between soil particles. The internal friction angle of the soil sample with higher vertical stress with a dry unit weight 17 kN/m^3 has decreased by about 8 degrees, and the sample with a dry unit weight 19 kN/m^3 has decreased by about 14 degrees.

As can be seen in **Figure 8**, the dirt filter grains are placed around the soil grains, causing the soil grains to move away from each other and lose them—roughness and friction of the soil.

Figure 9 shows the process of reducing the internal friction angle of soil contaminated with dirt filters on heavy and light vertical stresses. The reason for this reduction is because the polluting substance is placed between the soil grains, causing the grains to move away from each other. As a result, the internal friction angle of the soil decreases.

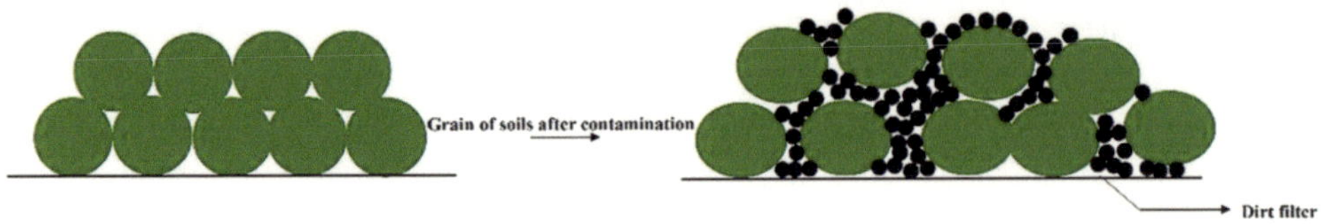

Figure 8.
Behavior of soil contaminated with dirt filter.

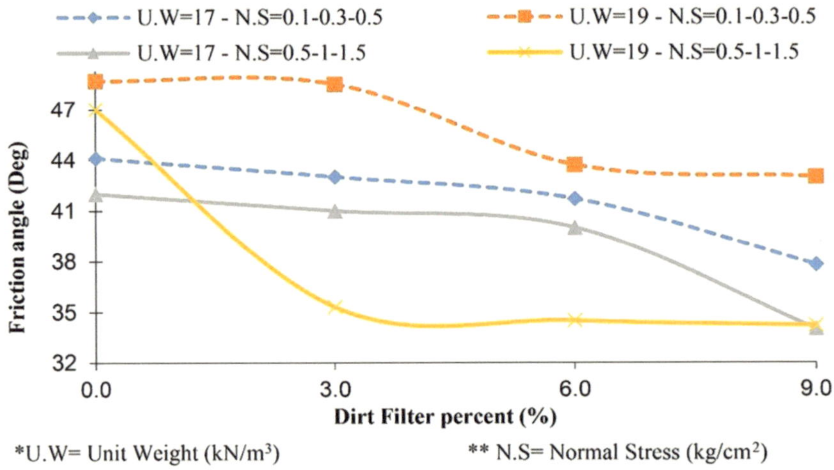

Figure 9.
Variation of internal friction angle of soil contaminated with different percentages of dirt filter.

3.2.2 The effect of contamination on the soil cohesion

By adding the polluting substance to the soil particles, the particles of this polluting substance are placed between the soil grains, and its surface active agents create a van der Waals bond with silica and other soil elements due to the non-polarity of the acidic sludge. The pollution of these bonds has increased, and the soil particles around the acid sludge particles begin to bond and finally stick to the particles, which results in an increase in soil cohesion.

By adding a dirt filter to the soil, the cohesion of the soil increases, which is due to the change in the nature of the soil. With increasing pollution, granular soil behaves like cohesion soil. This process of increasing cohesion is shown in **Table 6**.

Figures 10 and **11**, respectively, show the increase in cohesion of soil contaminated with acidic sludge and dirt filters under light and heavy stresses.

Table 7 shows the results for other contamination soils. The changes in the internal friction angle and cohesion of the contaminated soil generally decreased, depending on the type of soil and the pollutant. For granular soil, lubrication between the grains caused grain sliding and a decrease in the internal friction angle.

3.2.3 The effect of contamination on the shear strength of soil

Figure 12 shows that the shear strength of soil contaminated with acidic sludge, which is dependent on the combined effects of the internal friction angle and soil cohesion, had no clear and constant trend of change. In general, except at a vertical

Dry unit weight (kN/m^3)	Vertical stress (kg/cm^2)	Contaminant content			
		0	3	6	9
(a)					
17	0.1–0.3–0.5	0.051	0.166	0.119	0.12
	0.5–1–1.5	0.107	0.265	0.267	0.169
19	0.1–0.3–0.5	0.129	0.16	0.182	0.342
	0.5–1–1.5	0.06	0.393	0.364	0.5
(b)					
17	0.1–0.3–0.5	0.05	0.074	0.076	0.13
	0.5–1–1.5	0.11	0.135	0.21	0.263
19	0.1–0.3–0.5	0.13	0.225	0.167	0.192
	0.5–1–1.5	0.06	0.207	0.229	0.258

Table 6.
Soil cohesion resulting from (a) dirt filter contaminants, and (b) acidic sludge.

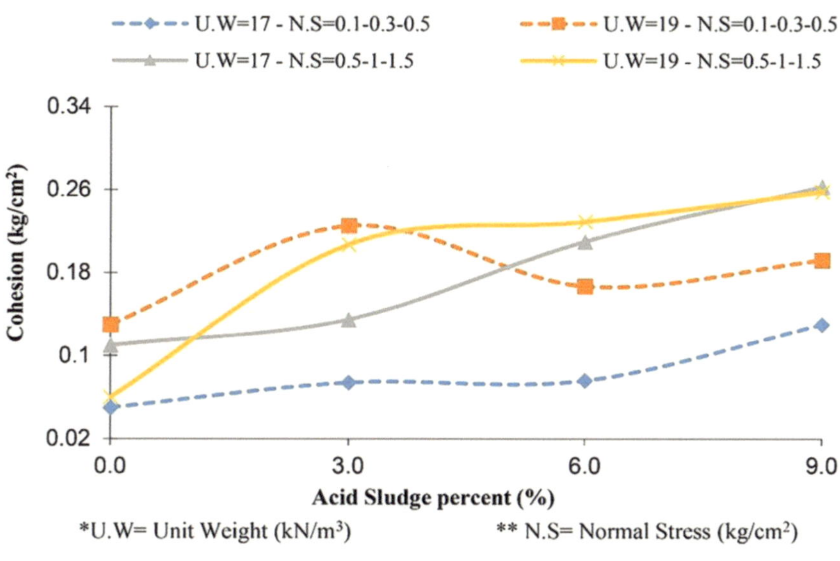

Figure 10.
Variation of cohesion of soil contaminated with different percentages of acidic sludge.

stress of 0.1 kg/cm^2, the shear resistance of the soil decreased. It is evident that, with an increase in the vertical stress of the soil, its shear resistance increased as the soil grains grew close together and the grains locked.

Low vertical stresses such as 0.1 kg/cm^2, contamination caused an increase in the shear resistance of the soil. At a higher vertical stress, the soil shear resistance decreased when contaminated with pollutants. The reason for this could be that, under a vertical stress of 0.1 kg/cm^2, the addition of a pollutant to the soil caused the filling of the space between the grains and eventually caused them to lock. This increased the shear resistance of the soil. Under vertical stresses, with the closeness of the grains, the addition of a pollutant increased the distance between the grains and

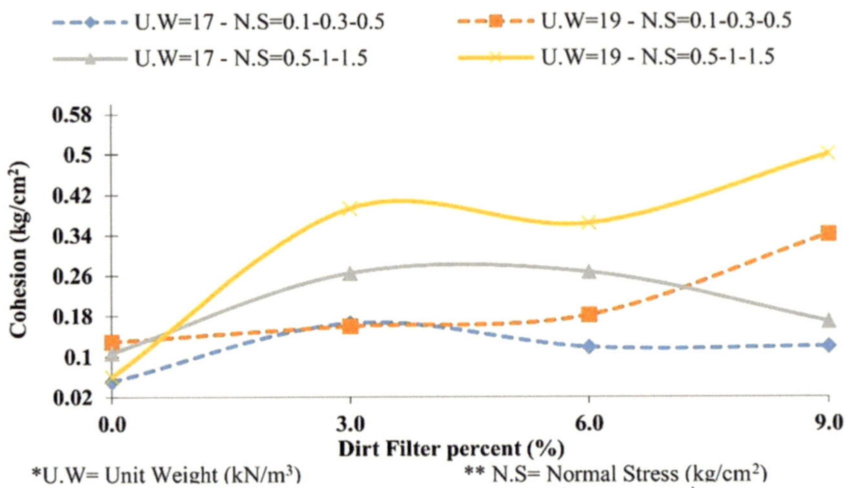

Figure 11.
Variation of cohesion of soil contaminated with different percentages of dirt filter.

References	Curing period	Vertical stress (kPa)	Content (%)	Type of contamination	Soil type	Soil friction angle (°)	Soil cohesion (kPa)
Safehian et al. [28]	A week	50–100–150	0	Gas oil	CH	39	52
			4			37	41.3
			8			36	34.5
			12			36	32.5
			16			35	31
			20			34	30.6
Khamehchiyan et al. [22]	One Month	55–110–160	0	Crude oil	SP	35	0
			4			32	3.3
			8			32	8
			12			29	7.6
			16			27	6.2
			0		SM	33	27.4
			4			33	19.9
			8			32	22.7
			12			26	20.8
			16			26	34
			0		CL	26	74.8
			4			27	28.4
			8			28	19.4
			12			29	19.4
			16			34	18

References	Curing period	Vertical stress (kPa)	Content (%)	Type of contamination	Soil type	Soil friction angle (°)	Soil cohesion (kPa)
Saberian and Khabiri [10]		55–111–222	0	Gas oil	SP	39	0
			3			35	10.2
			5.25			32	17.4
			8			31	26.7
Nasehi et al. [9]	One Month	24–44–64	0	Gas oil	CL	19	12.1
			3			14	11.4
			6			6	14.3
			9			6	14.6
			0		ML	21	9.9
			3			18	9.8
			6			14	11.3
			9			11	12.4
			0		SP	20	7.4
			3			19	8.4
			6			15	12.4
			9			14	12.2
Askarbioki et al. [29]	48 Hours	50–100–150	0	Gasoline	SC	28	14.9
			1			25	13.7
			3			23	13.5
			5			23	13
Rasheed et al. [30]	One Day		0	Gas oil	SPSM	12	18
			3			17	11
			5			26	4
			7.5			32	1.1
			0	Gasoline		12	18
			3			14	16.5
			5			15	14.2
			7.5			20	12.3
Karkush and Kareem [11]	Four Days	50–110–220	0	Crude oil	CL	37	66
			10			30	37
			20			25	22
Choura et al. [26]			0	Crude oil	SP	31	
			5			52	
			10			40	
Karpuzcu et al. [25]	Two Weeks	100–200–300	0	Crude oil	CL	26	0
			6			31	42.9
			12			23	67.3
			0		SC	39	24.4
			6			30	102.4
			12			36	139.5

References	Curing period	Vertical stress (kPa)	Content (%)	Type of contamination	Soil type	Soil friction angle (°)	Soil cohesion (kPa)
Sarmadi et al. [19]	Two Weeks	50–100–150–200–250	0	Kerosene	SW	38	0
			3			35	5.6
			6			33	6.4
			9			32	6.8
			12			31	7.1
			15			30	7.3
			10			20	41
			20			18	36
			30			18	32

Table 7.
Cohesion and internal friction angles of soil contaminated with hydrocarbons by maximum specific dry weight density.

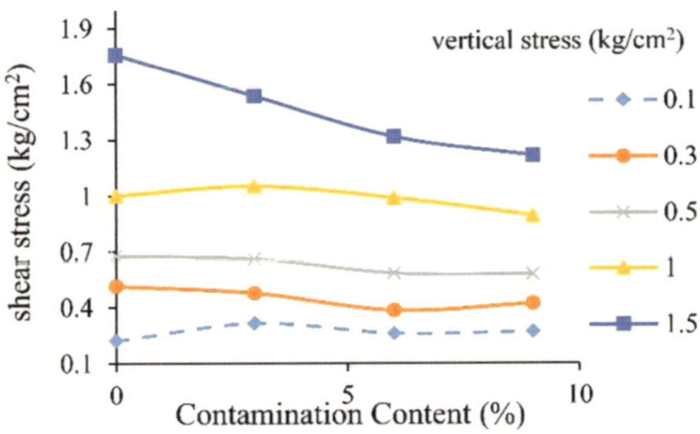

Figure 12.
Shear strength of soil contaminated with acidic sludge under vertical stress (0.1, 0.3, 0.5, 1, 1.5 kg/cm²) at a specific dry weight of 1.9 g/cm³.

made it easier for the soil grains to slide relative to one another. As a result, the shear resistance of the soil decreased.

Figure 13 shows that the shear strength of soil contaminated with dirt filter residue increased at all vertical stresses except 1.5 kg/cm^2. Under a vertical stress of 0.1 kg/cm^2, with the addition of 9% dirt filter residue to the soil, its shear strength increased from 0.221 to 0.479 kg/cm^2.

3.3 Soil vertical deformation

As shown in **Figures 12** and **13**, when the soil is contaminated, the vertical deformation does not show very different behavior in different percentages of contamination. In fact, in soil samples, there is a lot of conflict between the grains and the locking between them, and the grains cause expansion and expansion of the soil by moving on each other. In general, in most of the stresses, soil contamination has reduced the expansion of the soil (**Figures 14** and **15**).

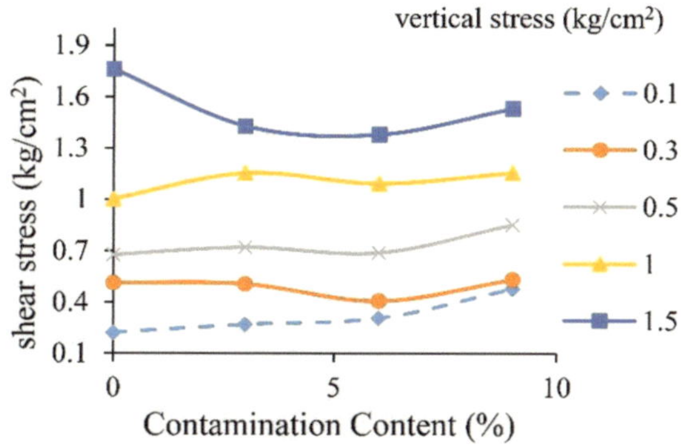

Figure 13.
Shear strength of soil contaminated with dirt filter under vertical stress (0.1, 0.3, 0.5, 1 and 1.5 kg/cm²) at specific dry weight of 1.9 g/cm³.

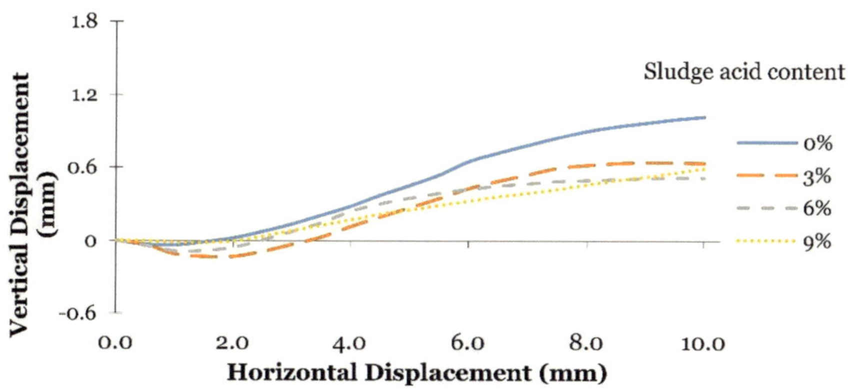

Figure 14.
Horizontal displacement-vertical displacement diagram of acid sludge samples with different contamination percentages under vertical stress of 1 kg/cm² and dry specific gravity of 1.7 g/cm³.

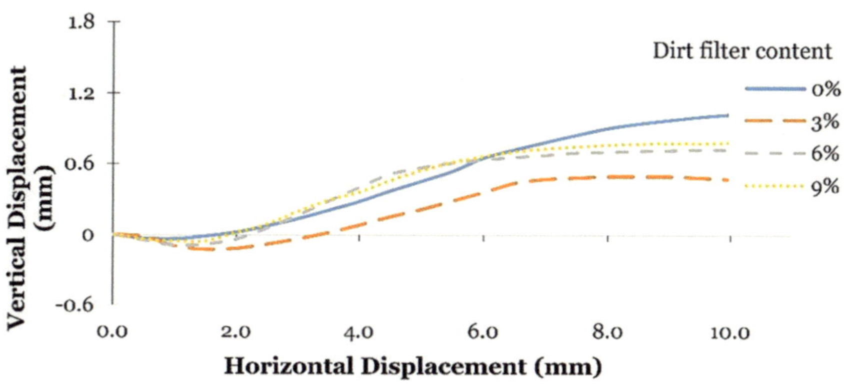

Figure 15.
Horizontal displacement-vertical displacement diagram of dirt filter samples with different contamination percentages under vertical stress of 1 kg/cm² and dry specific gravity of 1.7 g/cm³.

3.4 Effect of oil on soil structure

Clay particle placement can be categorized as mass or complex, coagulated or non-coagulated [31–33]. Therefore, the particles tend to be oriented face-to-face in a dispersed or divergent structure. If the forces are attractive, the particles tend to be oriented corner-to-face or corner-to-corner depending on the complex structure of the particles [34–36].

Considering that the structure of soil particles depends on the charge capacity of the electrostatic field, when clay is contaminated by hydrocarbons, their low dielectric constant and non-polarity will reduce the absorption capacity and force between the particles. This has a function similar to insulation; thus, it will affect the overall structure of the clay particles. As a result, the electrostatic field between the particles will push them toward repulsion and the degree of flocculation of the structure will decrease, giving it a scattered appearance [20].

Figure 16 shows that, with the addition of crude oil contamination to clay soil, the stickiness and low dielectric constant of the crude oil will cause a decrease in the double layer thickness that will cause edge-to-edge or edge-to-side formation of the soil particles. This means that, practically, the soil will become coagulated or flocculated.

Figure 17 shows that the addition of crude oil pollutants to a mixture of sandy and clay soil caused it to become flocculated with a scattered structure. The specific surface area of the soil decreased. This decreased the specific surface area of the soil and indicated a decrease in the resistance of the soil affected by the crude oil contamination [39].

In granular soil, unlike fine-grained soil, where the structure of the soil changes, the contaminant will surround the soil grains. The high viscosity of the hydrocarbon materials will cause lubrication between the soil grains and ultimately lead to a decrease in the surface tension between soil grains.

Figure 16.
SEM images of soil texture at 10,000× for uncontaminated samples and 20% crude oil contamination [37].

Figure 17.
SEM images of soil texture at 20000× of samples of: (a) 60% kaolinite +40% sand; (b) 60% kaolinite +40% sand +16% crude oil [38].

4. Discussion

Nowadays, the change of geotechnical characteristics of soil as a result of pollution by petroleum materials is considered as one of the challenges of geotechnical engineers. In this chapter, the effect of this pollution on the parameters of shear strength and shear behavior was investigated by performing direct shear tests and compression tests.

The results of the compaction test showed that soil contamination with hydrocarbon materials increases the compressibility of the soil. In fact, the high viscosity of hydrocarbon materials compared to water makes it easier for grains to move and slide more easily, which makes it possible to achieve higher densities.

The shear strength of the soil contaminated with hydrocarbon materials does not have a clear trend, but generally the internal friction angle of the soil decreases due to the easier sliding of the soil grains and the cohesion of the soil increases.

5. Conclusion

The following are the results of the present study

- the increase of acid sludge contaminations up to 4.2% due to the reduction of void soil and easier sliding of grains soil, the maximum dry soil unit weight increases.

- the dirt filter grains are placed around the soil grains, causing the soil grains to move away from each other and lose them—roughness and friction of the soil.

- with the addition of hydrocarbon materials to the soil, these materials are placed between the soil grains and cause a decrease in the firm connection between the soil particles and the easier sliding of the soil grains on each other, and finally, the internal friction angle of the soil is reduced.

- Under similar experimental and loading conditions, an increase in the contaminant content decreased the soil dilatancy angle. Under general conditions, the shear strength of soil will decrease with a decrease in the soil friction angle, which will decrease the dilatancy angle, except where the contaminant acts as a lubricant between soil particles.

- Dirt filter contamination at a dry unit weight of 1.7 g/cm^3 under high and low vertical strengths caused the contaminants to act as lubricants and boosted the shear strength of the soil. Further contamination filled the void spaces between the soil particles and caused a decreasing trend in soil shear strength.

6. Proposals for future research

In order to better analyze, suggestions for future research are presented as follows:

- Investigation of Shear Modulus variations of crude oil contaminated clay on small-strain range.

- The influence of hydrocarbon materials on the geotechnical properties of soil with Bending element, reinforcement column test.

- Designing a neural network in order to predict changes in geotechnical parameters of different soils using the results and data provided by other researchers.

- Investigating the effect of oil contamination on other geotechnical soil parameters including consolidation, Bearing capacity, settlement, permeability coefficient, liquefaction potential, and slope stability.

Author details

Mahdi Khodaparast* and Amir Khoshgoftar
Faculty of Engineering, Department of Civil Engineering, University of Qom, Iran

*Address all correspondence to: khodaparast@qom.ac.ir

References

[1] Aghajani S, Katebi H, Akhlagi T, Moradi GH. Crude Oil Spill Effect on Geotechnical Properties of the Tabriz Refinery's Soil. MSc Thesis. Tabriz-Iran (IN Persian): University of Tabriz Faculty of Civil Engineering Department of Geotechnical Engineering; 2017

[2] Riser-Roberts E. Remediation of Petroleum Contaminated Soils: Biological, Physical, and Chemical Processes. New York, United States of America: CRC Press; 1998

[3] Al-Rawas AA, Hago AW, Al-Sarmi H. Effect of lime, cement and Sarooj (artificial pozzolan) on the swelling potential of an expansive soil from Oman. Building and Environment. 2005;**40**:681-687. DOI: 10.1016/j.buildenv.2004.08.028

[4] Virkutyte J, Sillanpää M, Latostenmaa P. Electrokinetic soil remediation – Critical overview. Science of the Total Environment. 2002;**289**:97-121. DOI: 10.1016/S0048-9697(01)01027-0

[5] Al-Sanad HA, Eid WK, Ismael NF. Geotechnical properties of oil-contaminated Kuwaiti sand. Journal of Geotechnical Engineering. 1995;**121**(5):407-412. DOI: 10.1061/(ASCE)0733-9410(1995)121:5(407)

[6] Jukic A. Petroleum Refining and Petrochemical Processes Crude Oil: Composition, Classification. Croatia: Faculty of Chemical Engineering and Technology, University of Zagreb; 2013. pp. 37-46

[7] Cyrus S, Kumar T, Abraham B, Sridharan A, Jose B. Effect of industrial wastes on the physical and engineering properties of soils. In: Proceedings of Indian Geotechnical Conference GEOtrendz; 16-18 December 2010. India: IGS Mumbai Chapter & IIT Bombay; 2010. pp. 357-360

[8] Khoshgoftar A, Khodaparast M, Sedighi M. Effect of residues from a burnt oil refinery on the compaction parameters and strength of clayey sand. Bulletin of Engineering Geology and the Environment. 2021;**80**(8):6331-6341. DOI: 10.1007/s10064-021-02320-4

[9] Nasehi SA, Uromeihy A, Nikudel MR, Morsali A. Influence of gas oil contamination on geotechnical properties of fine and coarse-grained soils. Geotechnical and Geological Engineering. 2016;**34**(1):333-345. DOI: 10.1007/s10706-015-9948-7

[10] Saberian M, Khabiri MM. Effect of oil pollution on function ofsandy soils in protected deserts and investigation of their improvement guidelines (case study: Kalmand area, Iran). Environmental Geochemistry and Health. 2016;**40**(1):243-254. DOI: 10.1007/s10653-016-9897-y

[11] Karkush MO, Kareem ZA. Investigation of the impacts of fuel oil on the geotechnical properties of cohesive soil. Engineering Journal. 2017;**21**(4):127-137. DOI: 10.4186/ej.2017.21.4.127

[12] Kermani M, Ebadi T. The effect of oil contamination on the geotechnical properties of fine-grained soils. Soil and Sediment Contamination: An International Journal. 2012;**21**(5):655-671. DOI: 10.1080/15320383.2012.672486

[13] ASTM D698-12E2. Standard Test Methods for Laboratory Compaction Characteristics of Soil Using Standard

Effort (12 400 Ft-Lbf/ft3 (600 kN-M/m3)). West Conshohocken, PA: ASTM (American Society for Testing and Materials); 2007. DOI: 10.1520/D0698-12E02

[14] ASTM D3080. Standard Test Methods for Direct Shear Test of Soils under Consolidated Drained Conditions. West Conshohocken, PA: ASTM (American Society for Testing and Materials); 2003. DOI: 10.1520/D3080-03

[15] Shroff AV. Properties of oil contaminated soil and their remedial methods by admixtures-a case study. In: Indian Geotechnical Conferences. Vadodara, India; 1997. pp. 399-400

[16] Asadzadeh S, de Souza Filho CR. Characterization of microseepage-induced diagenetic changes in the upper-red formation, Qom region, Iran. Part I: Outcrop, geochemical, and remote sensing studies. Marine and Petroleum Geology. 2020;**117**:104-149. DOI: 10.1016/j.marpetgeo.2019.104149

[17] Nadri R, Mohgel M, Bahroudi A. Bidhand Strike-Slip Fault System (South Qom). Vol. 19. IN Persian: Earth Sciences; 2010. pp. 177-184

[18] ASTM (2003b). Standard Test Methods for Particle-Size Analysis of Soils, in D 422

[19] Sarmadi MS, Zohrevand P, Rezaee M. Effect of kerosene contamination on the physical and mechanical properties of sandy soil. Innovative Infrastructure Solutions. 2019;**4**:1-9. DOI: 10.1007/s41062-019-0196-1

[20] Fazeli G, Farrokhi F, Firoozfar A. Determination and Evaluation of Bearing Capacity of Clay Soil Contaminated with Oil Pollutants and Studying the Geotechnical Characteristics of the Substrate [Thesis]. Zanjan-Iran (IN Persian): University of Tabriz Faculty of Civil Engineering; 2019

[21] George S, Ea A, Sabu B, Np K, George M. (2014). Study on geotechnical properties of diesel oil contaminated soil. International Journal of Civil and Structural Engineering. 2014;2:113-117. DPI:10.13140/RG.2.2.17938.58564

[22] Khamehchiyan M, Charkhabi AH, Tajik M. Effects of crude oil contamination on geotechnical properties of clayey and sandy soils. Engineering Geology. 2007;**89**(3-4):220-229. DOI: 10.1016/j.enggeo.2006.10.009

[23] Roshanqhiyas M, Bagheripour M. Effect of crude oil contamination on the geotechnical properties of kaolinite clay soil in the road. Transportation Infrastructure Engineering. 2019;**5**:101-112. DOI: 10.22075/JTIE.2019.17368.1371

[24] Ghadyani M, Hamidi A, Hatambeigi M. Triaxial shear behaviour of oil contaminated clays. European Journal of Environmental and Civil Engineering. 2019;**23**(1):112-135. DOI: 10.1080/19648189.2016.1271359

[25] Karpuzcu M, Arbili M, Ghaffoori F, Cabalar A. An assessment of the effects of light crude oil on the geotechnical properties of a soil. International Journal of Advances in Mechanical and Civil Engineering. 2018;**5**:1-4

[26] Choura M, Salhi S, Cherif F. Mechanical behaviour study of soil polluted by crude oil: Case of Sidi El Itayem oilfield, Sfax, Tunisia. Environmental Earth Sciences. 2009;**59**(3):573-580. DOI: 10.1007/s12665-009-0055-z

[27] Karkush MO, Jihad AG. Studying the geotechnical properties of clayey

soil contaminated by kerosene. Key Engineering Materials. 2020;**857**:383-393. DOI: 10.4028/www.scientific.net/KEM.857.383

[28] Safehian H, Rajabi AM, Ghasemzadeh H. Effect of diesel-contamination on geotechnical properties of illite soil. Engineering Geology. 2018;**241**:55-63. DOI: 10.1016/j.enggeo.2018.04.020

[29] Askarbioki M, Kargaran Bafghi F, Mokhtari M, Khaleghi M. Impact of gasoline contamination on mechanical behavior of sandy clay soil. Journal of Mining and Environment. 2019;**10**(2):389-399. DOI: 10.22044/jme.2019.7660.1622

[30] Rasheed Z, Ahmed F, Jassim H. Effect of crude oil products on the geotechnical properties of soil. WIT Transactions on Ecology and the Environment. 2014;**186**:353-361. DOI: 10.2495/ESUS140301

[31] Lagaly G. H. van Olphen: An Introduction to Clay Colloid Chemistry. 2nd ed, John Wiley and Sons, New York, London, Sydney, Wiley Online Library. 1978.DOI:10.1002/bbpc.197800022

[32] Fakharnia BH, MA. and Khodaparast M. Performance of driven grouted micropiles: Full-scale field study. International Journal of Geomechanics. 2021;**21**(2):04020250. DOI: 10.1061/(ASCE)GM.1943-5622.0001905

[33] Mitchell jk. Fundamentals of Soil Behavior. New York: John Wiley and Sons, Inc.; 1976

[34] Hewayde E, Eng P, Abbas M, Kubba Z. Influence of engine oil on geotechnical properties of cohesive soil. International Journal of Engineering Research and Technology. 2019;**12**(1):33-41

[35] Mohammadi M, Khodaparast M, Rajabi AM. Effect of nano calcium carbonate (nano CaCO3) on the strength and consolidation properties of clayey sand soil. Road Materials and Pavement Design. 2021;**23**(10):2394-2415. DOI: 10.1080/14680629.2021.1976255

[36] Mohammadi M, Rajabi AM, Khodaparast M. Experimental and numerical evaluation of the effect of Nano calcium carbonate on geotechnical properties of clayey sand soil. KSCE Journal of Civil Engineering. 2022;**26**:35-46. DOI: 10.1007/s12205-021-1914-8

[37] Kangani M. Investigating the Mechanical Properties of Soils Contaminated with Crude Oil. MSc Thesis. Tehran-Iran (IN Persian): University of Water and Electricity Industry Faculty of Water and Environmental Engineering; 2012

[38] Ghadimi N. Investigation of the Effect of Organic Contaminants on the Strength Parameters of Sandy Clay Soil. MSc Thesis. Ardabil-Iran (IN Persian): University of University of Mohaghegh Ardabili Faculty of Civil Engineering; 2018

[39] Vaziri N, Ghafori M, Blouribazaz G. Investigating the Effect of the Presence of Diesel on the Atterberg Limit of Sandy Clay Soils. IN Persian: National Conference on Architecture, Civil Engineering, Urban Development and Horizons of Islamic Artin the Second Step Statement of the Revolution Tabriz Islamic Art University; 2020

Chapter 3

Soil Biogeochemical Factors Influencing Mobilization of Toxic Elements in Mine-Contaminated Soils and Remediation Potential of Plants

Albert Kobina Mensah and Emmanuel Amoakwah

Abstract

This study provides a current review on the impact and presence of potentially toxic elements (PTEs) in the environment as they pertain to mining operations. Emphasis is placed on the significance of implementing strategies to mitigate PTE contamination to safeguard the health of humans, plants, and animals. The significance of soil biogeochemical or governing factors that affect the mobilization of potentially toxic elements in mine-contaminated soils is examined in this study. Furthermore, these parameters assist the researcher or scientist in determining which remediation and mitigation strategies are most suitable for the contaminated site. The paper explains how soil pH impacts the toxicity and behavior of metal and nonmetal species. It describes how the mobility of cationic trace elements is enhanced at lower pH levels, whereas the mobility of anionic species is increased at higher pH levels. Additionally, the significance of soil redox chemistry in relation to the mobility and solubility of trace elements, specifically in the presence of inundation, is thoroughly elaborated in this review. Furthermore, this study presents the impact of redox conditions on the fare of transport of PTEs in flooded soils and aquatic environments. Ultimately, we argue compelling justifications for eco-friendly alternatives, revegetation, phyto-cleaning of PTEs, and restoration of contaminated mine sites.

Keywords: soil redox chemistry, sesquioxides, phytoremediation, potentially toxic elements, arsenopyrites, soil biogeochemistry

1. Introduction

Mining has serious consequences for the environment, causing soil and water quality to deteriorate as heavy metals accumulate. This has an unfavorable effect on the aquatic habitats of various fish species. The activity also has an impact on the volume of water supplied to mining towns, resulting in an increase in water treatment costs. This, in turn, has a negative impact on the economic well-being of those who live in these areas and rely on water supplies. According to Mensah et al. [1] and Mensah and Tuokuu [2], the long-term social and environmental effects of mining in host communities can be attributed to a number of flaws, including insufficient community participation or involvement of project-affected individuals, difficult and time-consuming registration processes for small-scale mining businesses, insufficient mining policies, and lax enforcement of mining regulations [4].

Potentially toxic elements (PTEs) like arsenic, cadmium, mercury, lead, antimony, selenium, and thallium can be found naturally in the environment but are also released by humans, notably in mining and industrial processes. These PTEs can contaminate the land, water, and air, causing serious health concerns to people and other species. High PTE concentrations can cause acute toxicity and even death, while low quantities can cause diseases and problems to develop over time. The presence of PTEs in mining sites is a big concern since it can contaminate nearby vegetation, crops, soil, surface water, and groundwater. Strategies for land restoration and alternative land use are required to reduce these dangers and protect human, plant, and animal health. Arsenic, for instance, is a well-known carcinogen that can result in a variety of cancers. Other PTEs, such as cadmium, mercury, and lead, have a negative impact on human health, disrupting vital enzymes, kidney function, and brain processes. Overall, the presence of PTEs in the environment poses serious health hazards, necessitating comprehensive management and remediation measures.

To ensure the mining industry's long-term sustainability, polluted and damaged mining sites must be gradually restored and rehabilitated during the active period and at the end of the mine's project life cycle. Furthermore, it is critical to apply sustainable measures to repair the damage caused to the land during mineral extraction, restore mine sites, and ensure the safe reutilization of the land. Inevitably, such measures must be implemented to preserve the survival of the ecosystem and the well-being of both humans and animals in mining areas. There is an urgent need for mild remediation studies to address the potential harm caused by contaminated abandoned gold mining spoils [3]. One of the strategies employed is "green remediation," which has been identified as a moderate *in situ* remediation strategy [4].

So far, a variety of physical and chemical techniques have been used or proposed to restore or recover the integrity of polluted and deteriorating mine sites. The use of physical and chemical stabilizing techniques in the restoration of mine-affected lands is a topic of ongoing discussion in scientific literature. These processes have the ability to either impede or facilitate the movement of potentially hazardous elements (PTEs) in contaminated areas [5, 6]. The physical strategy mostly involves filling excavated holes, modifying the topography, and then removing contaminated dirt from the surface and replacing it with clean, uncontaminated topsoil. Furthermore, the physical method uses containment techniques like barriers, caps, and liners to prevent dangerous substances from migrating and spreading into the surrounding ecology and mining communities [7].

Chemical treatments using oxides (e.g., iron oxides) have also been used to stabilize pollutants, reducing their solubility and thereby reducing their environmental and human health consequences [5, 8, 9]. Mensah et al. [3] found that using compost and manure on an abandoned mining site contaminated with As in Ghana increased the bioavailable As content by 332 and 315%, respectively. Hussain et al. [10] found that applying cow dung manure to an arsenic-contaminated area increased soil organic carbon content. This, in turn, reduced the availability of As in a paddy field in Pakistan. Alternatively, changes in soil pH and redox conditions may cause the dissolution of organic carbon and metal oxides used in chemical treatments. They may then release these compounds when PTEs adhere to or are absorbed by their surfaces, contaminating groundwater and the food chain [11–13]. Furthermore, Yang et al. [14] observed that applying biochar to a polluted area during floods increased the solubility of arsenic (As), resulting in higher mobilization and release of the contaminants.

Thus, the successful deployment of these physicochemical technologies to repair large areas of heavily polluted land may necessitate major financial investment as well as specialist skills in attentive monitoring, continuing quantification, and evaluation. At the same time, the viability of employing these approaches to rehabilitate substantial areas of polluted and deteriorating mining sites may fall short of the standards for socioeconomic and environmental sustainability for the mining corporation, local population, and adjacent ecosystems.

Phytoremediation through revegetation, which uses plants to restore polluted and damaged mine sites, could be a low-cost way for the mining industry to achieve long-term sustainability for the host mining community, its environment, and its economy. For example, the strategy could be used to address heavy metal contamination issues in gold mine soils [15, 16]. Therefore, by reducing food chain contamination, limiting the mobility of PTEs, reducing metal transfer into surface and groundwater, and guaranteeing a metal-contaminated-free environment, and mine land revegetation could lessen the negative effects of PTEs on human health. Furthermore, revegetation may aid with carbon sequestration, regulate wind and water erosion, guarantee water quality, lessen runoff from contaminated mining waste, preserve and improve soil fertility, and conserve soil moisture. In Ghana, revegetation is still a developing environmental engineering science that requires more in-depth investigation and study to expand its applicability. In addition, the pros and cons of various case studies, experiments done in the field and in a greenhouse, and applications in restoring mine-damaged sites from different mining companies in Ghana need to be documented and carefully evaluated. By so doing, the success stories and positive lessons can be upscaled and the challenges rectified to clean and reclaim the country's many polluted and degraded mine sites. To this end, we critically:

i. provide the most recent review and elucidation on the factors that could influence mobilization and bioavailability of PTEs; and

ii. appraise the science of revegetation for phytoremediation of PTE-contaminated mine sites.

2. Soil biogeochemical factors influencing bioavailability and mobilization of potentially toxic elements

2.1 Influence of soil pH

Soil pH is an important component influencing sorption and desorption, oxidation state, solubility, mobility, and therefore toxicity of metal and nonmetal species in soils [17–19]. Under acidic conditions, $H+$ ions displace metal cations from soil components' cation exchange complexes (CECs), causing metals to be liberated from sesquioxides and variable-charged clays where they have been chemisorbed (specific adsorption). Metal retention by soil organic matter is also worse at low pH, resulting in more accessible metal in the soil solution. As a result, numerous cationic PTEs, including as Cd, Cu, Hg, Ni, Pb, and Zn, become more soluble and mobile in soil solutions at low pH (below than 5.5). Thus, for cationic trace elements, lower pH results in more mobility and availability, whereas higher pH leads in greater mobility and availability for anionic species (e.g., [17]). For anionic PTEs, the argument is that as pH increases, the electronegativity of silicate secondary minerals decreases, facilitating availability. Furthermore, at high pH, the positive charge on the surfaces of oxides declines, as does their activity and efficacy for sorption, and thus, their retention for anionic PTEs decreases [20], whereas cationic characteristics rise. Many earlier investigations have revealed the impact of pH on PTE solubility and availability.

For example, Mamindy-Pajany et al. [21] stated that analyzing the behavior of As(V) on mineral adsorbents containing Fe (e.g., hematite and goethite) as a function of pH helps in determining the best treatment for As stabilization in polluted sediments. Furthermore, in their investigation on the mobility and phytoavailability of As in contaminated soil, Beiyuan et al. [22] found that pH is an important environmental component, with higher As mobility occurring at lower redox potential and higher pH. Furthermore, batch leaching tests at various pH levels by Al-Abed et al. [23] revealed a strong pH dependence on As and Fe leaching. This experiment aimed to better understand the effect of pH on As availability and solubility, as well as to explain pH as a significant element in adsorption and mobility.

According to Violante et al. [19], as pH increases within a specific range, soil sorption capacity reduces due to a decrease in the positive charge of minerals. Thus, increasing pH reduces the soil's sorption capacity for anionic PTE species [19]. Previous writers (e.g., [24–26]) interpreted anions' sorption capability in terms of "inner-sphere sorption and outer-sphere sorption" phenomena. They discovered that ions or molecules that are precisely sorbed replace OH- or OH2 groups on the surfaces of variable charge minerals. These reactions occur at low pH and cause OH- groups to receive protons. 'Inner-sphere sorption" is also known as particular sorption [26]. Arsenic species such as As(V) and As(III) form inner-sphere complexes [19]. O'Reilly et al. [27] pointed out that depending on pH of the medium and surface coverage, As(V) may form different surface complexes on inorganic soil components. For instance, As(V) is sorbed more than As(III) in a wide range of pH. Arsenite, on the other hand, is also reported to be sorbed more than arsenate in basic pH medium on ferrihydrite [19].

Mamindy-Pajany et al. [21] used adsorbate and adsorbent surface charge phenomena and ionization capabilities to explain the pH dependence of As species adsorption. They also said that negatively charged arsenate species ($H2AsO^{4-}$, $HAsO4^{2-}$, and $AsO4^{3-}$) predominate at higher pH. The surface charge of adsorbents is determined by proton transfer processes between the solution and the mineral surface, which can be positive, negative, or zero depending on the pH level. Iron oxides, for example, have a point of zero charge that corresponds to the pH value (pHZPC: pH at the zero point of charge), which means that the mineral surface charge is zero. In this regard, Mamindy-Pajany et al. [21] found that arsenate adsorption is facilitated when the surface charge of adsorbents is positive.

2.2 Influence of soil redox chemistry

Water saturation in sediments and soils affects their chemical and biological characteristics, microbial populations, and soil activities. Soils in such settings go through a series of reducing and oxidizing reactions when they transition from aerobic to anaerobic conditions during flooding [28]. Thus, the redox potential (EH) determines the solubility of anions and cationic PTEs during flooding conditions. This occurs either directly through changes in the soil EH caused by wetting and drying regimes or indirectly through changes in pH, dissolved organic carbon (DOC), and the chemical behavior of Al, Fe, Mn, and S [11, 17, 29, 30].

For example, a decrease in EH during flooding and rainy regimes may increase pH, hence improving As mobility, and vice versa [30, 31]. Also, redox potential could trigger metal reduction during reducing periods (e.g., from Al^{3+} to Al^{2+}, Fe^{3+} to Fe^{2+}, and Mn^{3+} to Mn^{2+}), resulting in the release of arsenic or mercury linked to reductively dissolved Al/Fe/Mn oxides. The opposite is true when oxidizing conditions are prolonged and dry.

Redox also influences PTE mobilization by inducing indirect changes in soil organic matter. Soil carbon is commonly considered as an excellent transporter of metals and nonmetals. This is because carbon's surface contains many positive ions, which provide fertile sorption grounds for PTEs and so reduce or increase their mobility. Under redox circumstances, organic matter has a wide range of effects on PTE mobility.

Mensah et al. [32] presented the following ideas for controlling or governing PTE mobility *via* the content of C or OM: (i) reductive dissolution of organic matter (OM) leading to the release of bound-PTE; (ii) pH change accompanying reduction reactions involving OM desorption—causes desorption of held-PTE; (iii) production of dissolved organic matter due to the presence of microbial biomass reduction of microbial biomass present in soil, which ultimately will affect PTE release; and (iv) reductive dissolution of organic matter leading to the release of Mn- and Fe-oxyhydroxides bound to organic matter.

The availability of potentially mobile PTE fractions influences the redox potential, which governs PTE's environmental toxicity. For example, potentially mobile fractions of As include un-specifically bound As, As specifically sorbed on mineral surfaces, As bound to amorphous and low crystalline iron oxides, and As bound to crystalline iron oxides [33, 34]. These fractions may become available due to environmental changes such as pH and redox conditions. For example, during reducing circumstances, As bound to Fe oxide fractions may be liberated [12], making it available for environmental pollution and accumulation in the food chain. Thus, flooding conditions and prevailing EH may influence plant uptake and translocation of critical elements and pollutants [35, 36].

Furthermore, redox circumstances have a significant impact on the fate of PTE such as As or Hg movement by changing their oxidation states. This impacts their mobility and toxicity in watery situations and waterlogged soils [37]. As previously stated, As(III) is more mobile than that of As(V). This could be aided by the presence of reducing chemicals in soil, which can reduce As(V) to As(III). Thus, the addition of organic matter or purposeful waterlogging of fields could hasten this decline and boost As availability [30, 38]. As(V) is extensively retained by inorganic soil components; hence, microbial oxidation causes As to be immobilized. Arsenic is present in well-drained soils as As(V), and in reduced soils as As(III), along with the elemental form As (As-0) and arsine (H2As) [39].

2.3 Influence of metal oxyhydroxides (Al/Fe/Mn)/sesquioxides

Metal oxides (Al, Fe, and Mn) have an important role in PTE mobility in soils and aquatic systems. Sesquioxides is the usual word for these metal oxides. Sesquioxides are oxides of an element or radical with a 2:3 ratio of the element's and oxygen's atom counts. Sesquioxides include oxides of phosphorus (III) and aluminum (Al2O3). Many sesquioxides, such as aluminum oxide Al_2O_3, lanthanum (III) oxide La_2O_3, and iron (III) oxide Fe_2O_3, contain a metal in the +3-oxidation state along with the oxide ion O_2^-.

Sesquioxides of iron and aluminum are found in soil. Alkali metal sesquioxides are unique in that they include both peroxide and superoxide ions. For instance, rubidium sesquioxides Rb_4O_6 are produced as $(Rb^+)_4(O_2^{-2})$ $(O^-2)_2$. Other exceptions to sesquioxides of metalloids and nonmetals include phosphorus (III) oxide P_4O_6, dinitrogen trioxide N_2O_3, and boron trioxide B_2O_3.

Metal oxides have high, active surface areas and are amphoteric, making them ideal for soil pollutant sorption and immobilization *via* specific sorption, coprecipitation, and the formation of inner- and outer-sphere complexes [8, 39]. Thus, the surfaces of iron hydrous oxides may play a crucial role in the preservation of PTE. A positive charged surface, for example, promotes As adsorption, and vice versa.

Metal oxides with large surface areas, such as ferric oxides, manganese oxides, aluminum oxides, titanium oxides, magnesium oxides, and cerium oxides, are regarded potential adsorbents for PTE treatment [40, 41]. Mench [42] found that applying iron oxides to As-contaminated garden soils resulted in decreased water extractable-As content and lower uptake in plant tissues.

Similarly, Hartley and Lepp [43] evaluated the efficacy of four Fe-bearing additions for As reduction in three contaminated soils (canal dredging, coal fly ash deposits, and low-level alkali waste). Goethite outperformed iron grit, iron II sulfate + lime, and iron III sulfate + lime in terms of reducing plant shoot As content.

Furthermore, the processes of protonation and deprotonation of metal oxyhydroxides or sesquioxides are dynamic mechanisms that have the ability to govern or manage the behavior of PTEs in polluted watersheds, floodplains, wetlands, and degraded mining regions. These mechanisms become obvious as the pH rises and the redox potential drops.

Metal oxyhydroxides are protonated in an acidic environment and deprotonated in a high pH environment. Surfaces hydrolyze when exposed to high pH levels, resulting in an increased presence of hydroxyl ions. This increase in hydroxyl ions reduces the surfaces' sorption capacity. In the aforementioned cases, cationic PTEs like lead, cadmium, copper, zinc, and other comparable elements show enhanced sorption. Such activities reduce the accessibility of toxic metals, reducing their potential to damage the surrounding ecosystem and, as a result, minimizing contamination throughout the food chain.

Furthermore, these characteristics increase the mobility and accessibility of anionic PTEs like arsenic and selenium. As a result, the surrounding ecology degrades and the food chain in these locations becomes more contaminated. Sesquioxide surfaces, particularly those consisting of aluminum, iron, and manganese, undergo protonation when exposed to lower pH conditions. These effects increase metal oxides' adsorption capacity for anionic PTEs while decreasing their ability to retain cationic PTEs.

As a result, protonation and deprotonation mechanisms are critical in determining the accessibility or absorption of PTEs in food crops growing in these areas. Furthermore, these systems control the flow of potentially hazardous elements (PTEs) into accessible surface and groundwater reservoirs.

2.4 Influence of soil organic matter

Soil carbon is primarily a transporter of PTEs. For example, it influences As (im)mobilization through a variety of mechanisms, including the formation of carbonates, carbon-As complexes, the presence of functional groups on carbon surfaces, the presence of positive and negative surface ions, indirect effects on medium pH, and the supply and activation of microbes required for As reduction and oxidation. For example, a lower pH reduces organic matter's negative surface charge, increasing its As adsorption ability, and vice versa.

Carbonates in calcareous soils or sediments act as an excellent buffer against pH drop [28]. Thus, the formation of carbonates from soluble organic carbon, such as during soil liming, may enhance the presence of negative charges while also facilitating the solubility and mobilization of As.

Organic matter indirectly influences the outcome of As transport and migration by serving as a food supply for bacteria. Furthermore, supply of C can stimulate microbial population and activity. This could catalyze a sequence of redox reactions in the presence of electron acceptors (such as oxygen and iron) [28]. These interactions significantly influence the mobility and availability of As in contaminated mining spoils. For example, adding wastewater to soils could increase carbon supply and nutrient content. As a result, crowded microorganisms may improve As reduction into more reduced forms [44]. Section 3 goes into greater detail about the effects of redox on PTE mobilization. 3. Humic acids released during organic matter decomposition affect the availability of PTEs significantly. The abundance of humic compounds in the native soil considerably influences the availability of PTEs. Humic acids with high molecular weight strongly bind PTEs, making them inaccessible in soil solutions and so limiting their release and availability for environmental pollution and food chain contamination. Low molecular weight humic acids, on the other hand, can chelate PTEs and prevent them from adsorbing on solid surfaces, increasing their mobility. Thus, low molecular weight humic acids are extremely active in soil and improve PTE binding abilities. Humic acids are significantly pH dependent, as shown by the following reaction:

$$R-COOH \leftrightarrow R-COO^{-} + H^{+} \tag{1}$$

where R is the carbon chain of organic matter.

This charge acquisition by organic matter occurs as a result of ionization, in which H+ is dissociated from or onto the active functional group, and the charge produced is highly dependent on soil pH. The negatively charged deprotonated form (R-COO-) is abundant and predominates at neutral to high soil pH, and it has the ability to adsorb cationic PTEs.

Heavy metals are soft acids, which can form organometallic complexes with organic matter's functional groups. The most active functional groups are carboxyl, alcohols, and phenols. As a result, the binding capability of the functional groups has a significant impact on metal availability. Dissolved organic carbon has been shown to reduce metal adsorption onto soil constituents by either competing for free metal ions and generating organometallic complexes or preferring adsorbed onto the solid phase. This phenomenon is more noticeable and effective at near-neutral soil pH because free metal ions are more readily available at high pH values.

The addition of organic matter to the soil enhances its Cation Exchange Capacity (CEC), allowing the soil to retain considerable amounts of cationic PTEs. As previously stated, this ability is largely dependent on soil pH. Humic acids develop a larger surface area at pH 6–8, and PTE retention peaks [45]. Humic acids' affinity for

PTEs rises in this pH range, controlling and influencing metal mobility. It should be noted that the stability of metal-ligand reduces as soil pH decreases, which supports R-COO-'s participation in metal complexation.

2.5 Influence of soil cation exchange capacity

The soil's cation exchange capacity is the ability to replace an ion on exchangeable sites with another ion having a higher or similar charge. This is caused by pH changes on the soil's colloidal surface or the displacement of one ion or charged particle by another as a result of ionic size differences. Because of its higher positive charge, aluminum has the ability to displace calcium from exchange sites. In some situations, magnesium, which has an ionic size similar to calcium, may substitute for each other on the exchange site. Thus, CEC controls soil ion binding capability [5].

The presence of positive or negative charges on the soil colloid influences its ability to restrict the flow of potentially harmful substances. The soil colloid serves as the location for soil reactions at the exchange site. For example, soils with a higher cation exchange capacity will be better able to retain cationic PTEs. Thus, soils with a high concentration of negative ions will be better able to inhibit the passage of positive or cationic PTE.

The kind of soil influences the mobility of PTEs and CECs. For example, Palansooriya et al. [5] found that soil containing clay minerals, metal oxides, and organic matter improves CEC and provides a larger surface area for PTE sorption, restricting PTE mobility. Soil organic matter, for example, consists of decomposed plant and animal wastes, as well as microbes. In addition, SOM contains humus or humic chemicals, which work synergistically to reduce PTE mobility at contaminated mine sites. Organic matter mostly retains PTE through ion exchange, complexation, and adsorption. Furthermore, the presence of phenols, carboxyl, carboxylate, and amino functional groups on soil organic matter surfaces provides PTE binding sites [5].

2.6 Effects of particle size distribution

PTE mobility is regulated by the distribution of soil particle sizes, which is dictated by the proportions of clay, silt, and sand. Soils with a higher percentage of clay contain more positive and negative ions. These ions help to regulate the release and liberation of potentially hazardous elements (PTEs) in contaminated watersheds, floodplains, agricultural areas, and mine sites. The clay fractions of the soil size distribution have larger concentrations of sesquioxides, notably Al, Mn, and Fe, than the coarse fractions, such as sand particles. Palansooriya et al. [5] support this position, reporting that soils with a high concentration of clay minerals have a stronger potential to adsorb chemicals due to their permanent or changeable charges, allowing them to retain more PTEs. Furthermore, the large surface area of clay particles allows for the absorption of both positively and negatively charged PTEs.

Furthermore, the clay fractions are predominantly composed of sulfides and have a higher surface area. The presence of sulfides controls arsenic release, as demonstrated by Mensah et al. [46] in their investigation of the consequences of arsenic contamination and its influence on human health in mine spoils in Ghana's southwest region.

It has previously been discovered that in extremely low oxygen conditions (i.e., decreased environments such as flooded and waterlogged habitats), sulfide formation is enhanced by the reduction of sulfate compounds, which results in the creation of Cd with sulfide minerals (i.e., CdS). According to Palansooriya et al. [5], this effect occurs in decreased paddy soils, where Cd solubility declines as CdS production increases. This occurrence highlights the relevance of sulfide in immobilizing Cd, particularly in paddy rice fields.

2.7 Effects of soil electrical conductivity (soil salinity)

Soil electrical conductivity (EC) is a chemical characteristic that influences the movement of PTEs in contaminated soils. Higher soil EC values are an indication of increased salinity and higher pH. High soil salinity is typically linked to an improvement in the soil exchangeable cations (Ca, K, Mg, and Na). These positive soil major elemental nutrients have the potential to compete with cationic PTEs for available adsorption sites [28]. Concomitantly, this may cause increased mobility of PTEs in soil with high salt content. Cadmium and other PTEs have a tendency to combine with chlorides in saline soil and cause in their mobility [47].

2.8 Effects of aging of the mine spoil

The age of contaminated sites and the duration of their abandonment and lack of reclamation both have a simultaneous impact on the accumulation, availability, and subsequent release of potentially toxic elements (PTEs). When compared to older soils, younger soils typically have a higher proportion of potentially mobile components of PTEs or also include a bigger quantity of mobile PTEs. When it comes to older soils, the proportions of PTEs that are either unavailable or residual generally take precedence.

Consequently, the age of the mine sites (whether active or abandoned spoils and tailings) that have been contaminated or degraded may consequently have an effect on the relative ease with which a metal or metalloid becomes available for release into the environment that is surrounding it, may contaminate the food chain, and may have an impact on the health of both animals and humans. To be more specific, it is projected that the residual percentage of the PTE that is present in the mining sites or the polluted mine spoils and tailings will contain a greater concentration of PTE.

Therefore, it is anticipated that an abandoned mine waste or residue that has not been reclaimed for a long period of time will slowly transfer PTEs from the accessible or available portion or the potentially mobile portion into the inaccessible portions. In this regard, Antoniadis et al. [17] showed that the PTE may be kept irreversibly onto interlayer soil sites over time, and that it can be hindered by lattice-fixed cations such as K or occluded by developed Al polymers.

In aged polluted mining sites, the phenomenon of soil pedogenesis will progressively relocate the elements from the accessible portion, leading them to become enclosed or immobilized by the crystalline. Consequently, the elements will gradually transition from primary minerals to secondary minerals. According to Shaheen and Rinklebe [48], an increase in the age of soil causes progression of pedogenesis and may result in an increase in the amount of crystalline Fe-oxides and hydroxides present in the soil. In addition, the ratio of Feo to Fed, which is also known as the activity ratio, can provide an indication of the age of the soil and show that there is a gradual shift towards crystallinity of Fe oxides with increasing age [48, 49]. Thus, it can be argued that older contaminated sites may be predominantly characterized by secondary minerals of elements. As a result, older mine spoil or tailings may have a higher concentration of unavailable forms of potentially toxic elements (PTEs), known as the residual fraction, compared to younger or currently active mining tailings or spoils.

This assertion is corroborated and supported by Mensah et al. [46], who found that secondary forms of arsenic, specifically scorodite, were prevalent in abandoned mine tailings, while primary forms were dominant in the active tailing in a study concerning arsenic contamination, potential mobilization and human health outcomes in southwestern Ghana. They further explained that the presence of scorodite in the mine tailings suggests that this secondary form of arsenic developed due to exposure to oxygen at the site. This happened due to oxidation and weathering of the primary mineral dominant at the site (i.e., arsenopyrite) [50, 51]. Similarly, Drewniak and Sklodowska [52] showed that scorodite is the most dominant secondary As mineral in mine-waste heaps and industrial deposits.

The site has been abandoned and left bare at the area over many years without any reclamation or provision of protective measures (e.g., vegetation cover). Von der Heyden [53] contended that these observations indicate that any remedial action implemented at such sites should target and focus on mitigating scorodite rather than arsenopyrites.

2.9 Influence of anions (chlorides, sulfate, phosphate, nitrate)

The chemistry of anions may greatly influence the release pattern and environmental ecotoxicology of PTEs from the mining spoil. Their chemistry also affects the choice soil remediation strategy for ameliorating contaminated and degraded mine sites. For instance, Mensah et al. [3] in a soil remediation incubation study found that higher contents of chloride were associated with manure and NPK fertilizer that were applied to a highly As-contaminated mine tailing. They explained that chloride may displace As-anions from sorption sites, particularly in manure and inorganic fertilizer enriched soils, improving their future uptake, availability, and release.

It has earlier been reported that the presence of chloride may affect the mobilization of As from the soil either *via* competition with As(V) or As (III) for available adsorption sites or by altering the electrostatic charge on soil minerals [31]. Generally, the presence of anions in the contaminated mine sites will cause mobilization PTEs and will restrict availability of positive PTEs. For instance, the presence of anions such as carbonates, chloride, phosphates, sulfates, and nitrate on the soil exchangeable sites may displace radical or ion of equal or lesser ionic size. Additionally, they increase the production of negative charges on the soil colloid and hence increase their sorption capacities for cationic PTES. On the other hand, their dominance on the exchange site facilitates the mobilization and hence the availability of metalloids.

3. Remediation of PTE-contaminated sites

Given the previously described environmental and human health risks posed by mining spoils to residents in mining towns, it is critical that gold mining be done in a sustainable manner. This involves the cleanup and repair of mine-contaminated and degraded areas. This section briefly discusses the many green and sustainable remediation solutions available for cleaning and lowering PTE contamination in mining lands. Recent and extensive assessments of these green technologies and sustainable solutions have been offered by Hou et al. [54], Palansooriya et al. [5], and

Wang et al. [9]. Bolan et al. [39] evaluated the literature on the remediation of trace element-contaminated soils and classified soil amendment approaches into mobilization and immobilization strategies.

3.1 Mobilization strategies for PTE remediation

3.1.1 Use of organic amendments

Organic soil materials are generally added to the soil to improve the soil physical, chemical, and biological properties. They thus improve the soil structure, improve the water holding capacity of soils, regulate soil temperature, regulate soil pH, improve the nutrients and organic matter status of the soil, as well as improve microbial population and activities. Consequently, soil amendments improve soil quality, encourage plant growth, and enhance crop performance. In other instances, organic soil amendments are added with the purpose of cleaning the soil and ameliorating presence of toxic elements in contaminated soil and water.

For instance, organic amendments such as sewage sludge (biosolids), manure, biochar, and composts can be added to the heavy metal-contaminated soils to increase mobility for subsequent uptake by plants. In this regard, Redman et al. [55] reported increases in leachable As from soils amended with compost. This was attributable to dissolved organic carbon (DOC) competition with As for sorption sites. Shaheen et al. [56] found that irrigation of soil with sewage effluent led to an increase in total As by 3.3 folds as compared to irrigation with clean water.

Furthermore, under flooding conditions, Yang et al. [14] discovered that pig-prepared biochar improved As solubility, which boosted its mobilization and release. They attributed this observation to two factors: (i). "anion exclusion" caused by the electrostatic repulsive force between negatively charged pig biochar and the dominant negative As species (e.g., H_2AsO_4, $H_2AsO_4^-$, $HAsO_4^{2-}$, and AsO_4^{3-}) and (ii). the production of soluble phosphate by the pig biochar, which competed with As anions for binding sites on soil particles.

3.1.2 Application of inorganic fertilizers and liming materials

Acidification of the soil has a tremendous effect of enhancing the mobility of trace elements in the soil [57]. This occurs at low pH values. Application of ammonium-based fertilizers has the tendency to decrease the pH of the soil. When ammonium-based fertilizers (e.g., $(NH_4)_2SO_4$) are applied to the soil, there are principally two fates of the NH_4^+ released.

Firstly, NH_4^+ undergoes the process of acidification through which H^+ ions are released. Nitrification is the conversion of NH_4^+ ions into nitrate ions through an intermediate process. The process is a 2-stepwise reaction.

$$2NH_4^+ + O_2 \xrightarrow{\text{Nitrosomonas and nitrococcus}} 2NO_2 + 2H_2O + 4H^+ \quad (2)$$

$$2NO^{2-} + O_2 \rightarrow 2NO_3^- \quad (3)$$

The release of H^+ ions in the first reaction causes an increase in soil acidity, and therefore, the pH of the soil is lowered. Under this acidic condition, the solubility of heavy metals is enhanced, and they are made available to be taken up by plants. In this scenario, trace elements or PTEs have increased capacity to be absorbed by plants and consequently results in contamination of the food chain. Pueyo et al. [58] showed that the use of ammonium-based fertilizers increases the solubility of heavy metals such as Cu by creating strong amino complexes.

Secondly, the NH_4^+ ion released after the application of ammonium-based fertilizers can also be taken up by plants directly. In this case, H^+ is released by the plants upon the uptake of NH4$^+$, and this H^+ released also lowers the pH of the soil. More importantly, nitrogen is a major element needed by plants for their physiological functions. Therefore, application of N fertilizers is important to help improve the nutritional status of the soil medium, and also, it contributes immensely to plant nutrition. In this case, the plants produce a significant amount of biomass to accumulate elevated levels of heavy metals. This scenario increases the uptake and phytoavailability potential of hyperaccumulators.

Inorganic amendment like $CaCl_2$ is also known to facilitate the uptake of heavy metals by plants. Most experiments that critically examine the mobility of metals using salt as an amendment to enhance phytoextraction focus on the application of Cl^- salts, more especially $CaCl_2$. The application of chlorides results in the formation of complexes of heavy metals with the chloride anion, for instance, $CdCl^+$ complexes. Application of salt ($CaCl_2$) to the soil results in the dissociation of the salt to positively charged and negatively charged ions. The cationic component has the propensity to undergo an exchange reaction with heavy metals on the sorption sites, and Cl-metal complexes are formed, which are soluble. Concurrently, the divalent cation Ca^{2+} can remove the already adsorbed metals from the exchange complex into the soil solution; hence, the metals become bioavailable. McLaughlin et al. [59] reported a significant increase in the uptake of Cd by tubers of *Solanum tuberosom* after the application of chloride.

Application of P fertilizers to As-contaminated soils can increase As availability and consequent environmental health threats. The role of P in As fate of transport and release has been extensively explained in previous sections. Thus, P fertilizers may be used to enhance the phytoextraction efficiency of hyperaccumulating plants for As.

Liming can also be done to increase and enhance mobilization strategies for As removal. Generally, liming is carried out to correct the soil pH, by decreasing the presence of protons and increasing the presence of hydroxyl ions on the soil colloids. Liming materials, for example, carbonates, oxides, and hydroxides of Ca, Mg, and Na, can be used to raise the soil pH and reduce acidity. Also, eggshells, mussel shells, and oyster shells, biochar, red mud, and clay minerals have soil liming abilities [5].

Calcite ($CaCO_3$) at high soil pH values has the tendency to sorb Cd as $CdCo_3$, and this makes Cd unavailable. Other metals also succumb to sorption by calcite and hence are made unavailable, and this possibly explains why acidic soils are limed to reduce the availability of metals to plants. The affinity of calcite for metals is related to their ionic strength. This is more evident for Cd^{2+} that, at high pH, substitutes for Ca^{2+} by chemisorption because the radius of Cd fits better for this process [45].

Applying higher amounts of MgO may result in the mobilization of anionic PTEs, such as As, and immobilization of beneficial micronutrients such as Fe, Mn, and Zn [60]. It is thus recommended that determining suitable soil qualities and predicting expected soil pH are key things to consider prior to addition of lime [5]. Nonetheless, applying liming materials together with other soil amendments can maximize liming efficiency and reduce the associated-environmental effects. Higher pH, reduction in acidity, production of more negative ions and charges, and increased formation of total CO_3^{2-} associated with liming are key reasons underpinning their influence on As mobility.

3.2 Immobilization strategies for PTE remediation

Immobilization of As could also be achieved chiefly through adsorption, precipitation, and complexation reactions that result in the redistribution of contaminants from solution phase to solid phase and consequently reduce their bioavailability. Examples include clay, cement, zeolites, minerals, phosphates, organic composts, and microbes [39]. Some studies (e.g., [43, 61]) have indicated the potential of residues such as red mud and iron oxides for treating As-contaminated sites and reducing their bioavailability. Others have combined treatments, such as modified red mud or iron oxide, with biochar to reduce As bioavailability and release into the environment [62]. Most immobilization technologies may be performed *ex situ* or *in situ*. *In situ* processes are preferred due to lower labor and energy requirements, but implementation of *in situ* will depend on specific site conditions [7]. The overall method for solidification/stabilization treatment operations is to combine or inject treatment ingredients into contaminated soils [7].

3.3 Plant techniques for PTE remediation

This method involves the use of plants to facilitate removal of metal and metalloid contaminants from the soil matrix [63–65]. According to Jiang et al. [66], phytoremediation, a green approach using plants to remediate toxic compounds, is a cost-effective, socially acceptable, and environmentally friendly technology for soil and groundwater cleanup. Reeves et al. [65] claimed that hyperaccumulator plants have concentrated on the prospect of removing significant amounts of the element from the soil while using standard fertilizers. They added that the rate of biomass production and the concentration of the desired element that can be achieved in harvestable plant matter are critical factors to be considered when selecting species for phytoremediation. Other factors to consider in selecting suitable species for mine land remediation are provided by other authors [63, 67, 68]. It is thus recommended that hyperaccumulation plants are those plant species that are capable of accumulating 1000 mg As/kg in the dry matter of any aboveground tissue [69, 70]. Recently, Mensah et al. [63] have observed that *Chromolaena odorata*, an indigenous plant near an abandoned mining spoil in Ghana, could offer potential for cleaning As from mining sites. In that study, *Chromolaena odorata* had As translocation factor of 4.7, further implying its ability to accumulate As from the mining soils. Translocation factor, bioconcentration factor, and bioaccumulation factors are other indices used to appraise the phytoremediation potential of plant species.

3.4 Revegetation of degraded and contaminated mined soils

According to Blum [71], the conventional approach to remediating degraded soils involves implementing extended periods of forest fallowing in a rotational manner. The composition and density of the soil have a direct impact on the long-term stability of the recovered plant community [72]. Soil serves as the fundamental basis for this process. In order to restore land to its original state and maintain a self-sustaining ecosystem, reclamation strategies should encompass various aspects such as soil structure, soil fertility, microbial populations, topsoil management, and nutrient cycling [72]. The reclamation and revegetation of abandoned mined areas are frequently constrained by several physical and chemical characteristics found in the soil. These characteristics encompass, but are not restricted to, low pH levels, elevated metal and metalloid concentrations (including metal salts), insufficient nutrient levels (e.g., N, P, and K), and inadequate or absent soil structure.

Mining activities can lead to the degradation or loss of soil structure and functions, resulting in soil toxicity, limited nutrient availability, and unfavorable soil texture. The form and function of soil, while constituting only a component of a larger ecosystem, serve as a microcosm that reflects the characteristics and dynamics of the entire ecosystem. According to Bradshaw [73], if these issues are not addressed, the re-establishment of vegetation and the restoration of ecosystem function may prove to be challenging or unattainable.

In order to enhance the effectiveness of rehabilitation and reforestation efforts, it is imperative to conduct studies and analyses on the biological and physiological traits of regenerated or newly planted trees, as well as the underlying mechanisms that impact their productivity [74].

In laboratory settings, mulches have been observed to enhance soil moisture conditions. However, it is important to note that achieving successful revegetation in real-world environments necessitates the elimination of salts from the root zone, as highlighted by Grigg et al. [75]. The presence of high salinity is a significant constraint for the establishment of plants, as it hampers the process of seedling emergence by delaying germination [76]. Consequently, this leads to an extended period during which the growth medium has to retain moisture [77]. Additionally, it hinders the growth of plants by inducing osmotic stress, hence exacerbating the issue of low moisture availability from precipitation [75].

The management of topsoil plays a crucial role in reclamation plans as it serves to mitigate nitrogen losses while simultaneously enhancing soil nutrient levels and microbial activity. According to Sheoran et al. [72], revegetation is well recognized and considered to be an effective method for mitigating erosion and safeguarding soil quality during the reclamation process. Ecological restoration and mine reclamation have emerged as integral components of the sustainable development agenda in numerous nations. Sheoran et al. [72] add that effective planning and environmental management practices can significantly mitigate the environmental consequences associated with mining activities, hence contributing to the preservation of ecological variety.

When endeavoring to rehabilitate a native ecosystem, the initial revegetation attempt is unlikely to provide vegetation that is identical to the original. The primary objective of the initial revegetation effort is to develop the fundamental elements necessary for a self-sustaining system, ensuring that the ideal vegetation complex is achieved through successional processes (see Sheoran et al., [72]). The optimal period for initiating vegetation growth is contingent upon the temporal pattern and consistency of precipitation throughout the seasons. It is imperative to ensure that all necessary preparation tasks are carried out prior to the period when seeds are most likely to encounter the requisite conditions for germination and viability, namely, consistent rainfall and appropriate temperatures [72].

Coppin et al. [78] provide a definition of revegetation as the systematic undertaking of vegetation establishment and subsequent maintenance, which is carried out as an integral component of reclamation, rehabilitation, or restoration efforts. One example of land reclamation is the filling of surface mines, followed by the recontouring of the land and the establishment of a vegetative cover that serves to safeguard the soil [79].

The primary emphasis of mine restoration endeavors has been directed towards nitrogen-fixing species belonging to the legume family, as well as various grasses, herbs, and trees. According to Sheoran et al. [72], plants that possess the ability to tolerate metals have proven to be efficacious in soils that are both acidic and contain high concentrations of heavy metals. Wong [80] argues that the restoration of plant cover on overburden dumps can effectively achieve several objectives, including stabilization, pollution control, visual enhancement, and mitigation of risks to human populations.

Phytoextraction, which involves the use of green plants for remediation purposes, shows promise as a feasible approach for the decontamination of extensive soil areas. Furthermore, it has been suggested that this method could potentially serve as a viable alternative to existing soil clean-up techniques [81–85].

The presence of vegetation has a crucial function in safeguarding the soil surface from erosion and facilitating the formation of fine particles [86, 87]. The degradation process can be reversed through the stabilization of soils, achieved by the growth of vast root systems. After being established, plants have the ability to enhance soil organic matter, reduce soil bulk density, regulate soil pH, and facilitate the upward movement and accumulation of mineral nutrients in an accessible state.

The root systems of these organisms enable them to function as scavengers of nutrients that are not easily accessible. The nutrients are accumulated by plants and subsequently redeposited onto the soil surface in the form of organic matter. This organic matter facilitates the availability of nutrients *via* microbial decomposition [87–90].

4. Revegetation in mitigating soil pollution with potentially toxic elements

According to Mensah et al. [46] and Ramírez et al. [91], mine wastes are important sources of contamination and point-sources for potentially toxic elements (PTEs) in soil, including Al, As, Cd, Cr, Cu, Fe, Mn, Ni, Pb, Ti, V, and Zn. PTEs may be transferred into surrounding plants, food crops, soil, surface water, and plants as a result of these. Metal pollution in soils and sediments may have long-term health effects on people and ecosystems by affecting aquatic life, food safety, and water quality.

In other situations, children who frequently use metal-contaminated terrain or abandoned mine sites as playgrounds run the danger of ingesting metal through their diet [92, 93]. Furthermore, as noted by Mensah et al. [46], communities may pass through unfenced abandoned mine spoil sites and breathe in polluted dust or consume contaminated mine wastes. In areas with gold mining, oral consumption of soil PTEs continues to be a significant exposure pathway that affects human health [46, 93, 94]. This could happen accidentally by ingesting dust, dirt, and soil, or purposefully by engaging in geophagy, which is the deliberate consumption of soil or earth materials like clay [95]. In Latin America and Africa, pregnant women frequently engage in the practice of geophagy [95, 96]. As is known to be present in metal ores, tailing deposits, soil, and dust in gold mining sites, raising serious health risks for both humans and animals [96].

For example, both acute and long-term exposure to As may result in health issues for humans, including skin damage, symptoms of cancer, and abnormalities in the circulatory system [97, 98]. Moreover, Wilson disease, which impairs the function of several organ systems, including the liver and the brain, is brought on by an excessive buildup of copper in the body as a result of consuming tainted food and water [99].

Furthermore, health issues like reproductive, developmental, hepatic, hematological, and immunological health defects; abortions, infertility; birth defects; and malformations may be linked to exposure to and accumulation of Cd, Cr, Pb, Ni, and V in essential body organs [100].

However, the positive environmental effects on ecological integrity and human health resulting from improper management of mine wastes [101], runoff from stockpiles [1], and abandonment of mine spoils left untreated [46] often outweigh the benefits associated with goldmining in Ghana, such as the creation of jobs and government revenues. These activities worsen the anthropogenic release and migration of PTEs into the environment, which has an impact on people's livelihoods and health.

Remediation of metal contamination issues in gold mine spoils has primarily involved soil washing and flushing [9, 39], removal and replacement of contaminated soil with clean soil [102], and chemical stabilization using oxides to reduce metal solubility [5, 8, 9]. In some cases, containment techniques have been used to lessen the migration and transfer of harmful components. These techniques include the use of built barriers, caps, and liners [7].

The two main drawbacks of these techniques are their high cost to many mining corporations and their difficulty level. Furthermore, PTEs cannot decompose like organic chemicals do; thus, the cleanup usually involves either removing them [15] or employing stabilization and immobilization strategies to reduce their uptake, surface runoff, or leaching [5, 39].

In order to address the issue of heavy metal contamination in gold mine soils, phytoremediation—the use of native plants—is suggested as a practical and affordable option to environmental management [15, 16]. When selecting potential plants for contaminated site phytoremediation, some authors (e.g., [67, 103–105]) have recently suggested that it may be wise to use well-adapted native species of the locality under investigation. Additionally, native and indigenous species are favored over foreign ones since they may be more suited to the local environment and fully integrate into the functioning ecosystem [68].

According to Ashraf et al. [106], phytoremediation is still a young field of study that requires further in-depth research to fully understand its applications. For example, the potential of native plants like *Pueraria montana, Alchornea cordifolia, Lantana camara,* and *Pityrogramma calomelanos*-fern [102, 104 107, 108] to clean mine-contaminated spoils has been investigated by Mensah et al. [63]. There is a potential of cleaning up these PTEs with native plant species. *Chromolaena odorata* (TF for As = 4.7, Cu = 5.1, Ti = 1.4, Zn = 2.0) and fern (TF for As = 1.50, Cu = 1.4, Ti = 1.8, Zn = 2.1) can be used to achieve phytoextraction. Their greater TFs for these elements suggest that *Chromolaena odorata* would be a better choice than fern. Furthermore, *Alchornea cordifolia* and *Pueraria montana* can be utilized for Cu, Ti, and Zn phytoremediation due to their TFs above 1. For Cu and Zn, *lantana camara* could potentially be a possibility. As a result, native plants like *Chromolaena odorata* and *Pityrogramma calomelanos*-fern can extract As, Cu, Ti, and Zn from multielement-contaminated mine soils and farms.

In Cd-contaminated soils in China, Liu et al. [107] investigated *Lantana camara* exclusively for Cd remediation and discovered that the plant exhibited high Cd tolerance, with bioaccumulation and translocation values more than 1. Furthermore, *Pteris vittata* L. showed a high potential for Pb phytoextraction among the nine native plants that Ashraf et al. [106] investigated for metal cleanup of contaminated mine tailings in Malaysia. Thus, by reducing food chain contamination, limiting their mobility, reducing metal transfer into surface and groundwater, and guaranteeing a metal-contamination-free environment, phytoremediation lessens the negative effects of PTEs on human health. Furthermore, phytoremediation may help with carbon sequestration, regulate wind and water erosion, guarantee water quality, lessen discharge from tainted mining waste, preserve and improve soil fertility, and save soil moisture.

In Mensah et al. [32], it was shown that the incorporation of manure and iron oxide (MFE) resulted in the phyto-stabilization of Co, Mn, Hg, Mo, Ni, and Zn in ryegrass, as evidenced by bioconcentration factor (BCF) values exceeding 1. The results of the study indicated that the effectiveness of ryegrass phyto-stabilization was significantly improved with the combined application of manure and iron oxide. Therefore, it was inferred that the simultaneous utilization of manure and oxides can be employed on mine tailings to enhance the growth of ryegrass and mitigate the release of potentially toxic elements (PTEs) into the surrounding environment.

Mensah et al. [3] conducted a 60-day pot experiment to explore the potential of ryegrass (*Lolium perenne*) along with different soil organic and inorganic amendments in phyto-cleaning. Arsenic in a heavily contaminated gold mining spoil in southwest Ghana. They found that ryegrass might be more useful for phytostabilization as its roots concentrate more As than its shoot. This was further supported by higher BCF above 1 as well as lower BAC and TF < 1.

4.1 Identifying appropriate phytoremediation technology

4.1.1 Direct and assisted phytoremediation

Phytoremediation can be classified into two distinct categories, namely, direct and indirect. The process of indirect or plant-assisted remediation entails the cooperative interaction between plants and a method of soil improvement in order to alleviate the existence of pollutants in soil that has been subjected to contamination.

4.1.2 Indirect or assisted phytoremediation

Soil amelioration encompasses the process of enhancing soil quality through the incorporation or application of various organic or inorganic soil supplements. These amendments may include substances such as biochar, compost, manure, iron oxide, inorganic fertilizers, sewage sludge, red mud, and other similar materials. These adjustments may be included alone or in conjunction with one another.

Pot experiments can be applied in evaluating and choosing an appropriate revegetation scheme for restoring a degraded mining site. According to Tordoff et al. [86], large-scale revegetation methods can first be tested in small-scale glasshouse pot experiments, followed by larger field trial programs. In this manner, the principal constraints to plant growth are identified, and appropriate corrective therapy can be selected and planned [109].

Overall, pot experiments in soil science offer a controlled and flexible platform for investigating a wide range of research questions related to soil properties, plant growth, nutrient management, and environmental impacts. The findings from these experiments contribute to the development of sustainable agriculture practices and environmental management strategies.

In a study conducted by Addai et al. [110], it was observed that the application of rice husk biochar, poultry litter compost, and NPK, either individually or in combination, to a gold mine tailing contaminated soil resulted in significant improvements in various soil properties. These improvements included an increase in soil pH, organic matter content, and available phosphorus, as well as an enhancement in the cation exchange capacity (CEC) with respect to calcium, potassium, magnesium, and sodium ions. Additionally, the application of these amendments led to a reduction in the levels of exchangeable acidity within the mine tailings.

Furthermore, using tailings enhanced with nitrogen, phosphorus, and potassium (NPK) resulted in a significant 42% increase in nitrogen uptake. Furthermore, applying biochar and NPK to tailings increased plant uptake of phosphorus (P) and potassium (K) by 128 and 101%, respectively. In terms of PTE uptake, specifically Pb, Hg, As, Cd, and Cr, the use of biochar, poultry litter compost, and their combinations resulted in a reduction of Hg uptake (measured in mg/kg) in the pot by up to 49% when compared to the control group.

The ability of plants to absorb high amounts of metal in a short period of time has a significant impact on phytoremediation efficiency. Hyperaccumulators acquire significant amounts of metal in their tissue independent of metal concentration in the soil [111], as long as the metal is there.

This is in contrast to the moderate accumulators that are now being used for phytoextraction, where the amount of absorbed metal is proportional to the concentration of the soil. Although the total metal content of the soil is substantial, the fraction easily available in the soil solution controls the effectiveness of metal absorption by plant roots. To improve the speed and quantity of metal removal by plants, some studies propose the use of various compounds to increase the amount of accessible metal for plant uptake. Acidifying agents [112, 113], fertilizer salts [114], and chelating compounds [81, 83] are among the chemicals proposed for this purpose. These compounds increase the quantity of bioavailable metal in soil solution by either freeing or displacing metal from the soil's solid phase or making precipitated metal species more soluble.

Research in this area has been relatively successful, but the prudence of releasing huge amounts of harmful metal into soil water is debatable. The availability of heavy metals in the soil solution, which enhances plant uptake, is critical to the efficacy of phytoextraction. Several adjustments have been recommended to increase the availability of heavy metals for successful phytoextraction research.

4.1.3 Direct or unaided phytoremediation

In contrast, direct phytoremediation refers to the process in which plants directly or naturally absorb and break down contaminants within their tissues. This approach may entail the direct establishment of vegetation on the degraded soil without the aid or involvement of soil amendments. Typically, this classification of natural rejuvenation entails the utilization of native or indigenous plant species that are adapted to the prevailing climate and soil conditions in the specific geographic area.

To obtain a comprehensive understanding of species selection and potential survival within degraded mining spoil, it is advisable for the project officer or environmentalist to conduct an observation of the immediate vicinity, such as the abandoned mine tailings. The identification of species thriving at the periphery of a damaged ecosystem or mine spoil provides insight into whether those species have the potential to endure the adverse conditions of the deteriorated or nutrient-depleted mining site.

4.2 Plant elemental uptake, phytoremediation potential, and identifying plant species for phytoremediation

The determination of plant element uptake and phytoremediation potential involves assessing the ability of certain plant species to absorb and accumulate contaminants from the soil, water, or air, with the goal of remediating polluted environments. This process helps in understanding which plants are effective at phytoextraction (removing contaminants from the soil) and phytostabilization (reducing the mobility of contaminants) and can be used to guide phytoremediation strategies.

To ensure the successful identification of appropriate plant species for phytoremediation endeavors, it is crucial to take into account the efficacy of metal accumulation. Plants can be categorized according to the degree of metal accumulation they exhibit, as well as the specific plant organ in which the accumulation occurs. The evaluation of metal accumulation in plants entails the examination of trace element concentrations in various plant organs, including shoots, roots, stems, fruits, seeds, and other plant components.

The assessment of a plant species' ability to tolerate and accumulate metal in their tissues is conducted by comparing concentration values with both normal and phytotoxic quantities. The assessment of a plant's potential to accumulate elements is conducted by comparing its accumulation levels with the established threshold criterion. These calculations can be applied to phytoremediation experimental trials and projects carried out in either the greenhouse or in the field. They can similarly be applied to studying and quantifying bio-accessibility and toxicity of metals in food crops that are grown in contaminated areas.

The determination of the phytoremediation capacity of the chosen plant species can be achieved by considering two primary parameters or indices:

i. Soil-to-plant transfer factor and

ii. Root-to-shoot plant transfer factor

4.2.1 Soil-to-plant transfer factors

The bioconcentration factor (BCF) and bioaccumulation coefficient (BAC) are used to evaluate the accumulation of trace elements (PTE) in plant roots and shoots. BCF calculates the ratio of element concentration in plant roots to soil content (Eq. (5)), whereas BAC computes the ratio of element concentration in plant shoots to soil content. These criteria evaluate a plant species' capacity to extract PTE from polluted soil. Fitz and Wenzel [115], Yoon et al. [116], and Shaheen and Rinklebe [48] all found that plants with BCF/BAC values less than one are not suitable for phytoextraction.

$$BAC = \frac{PTE\,content\,in\,shoot}{Total\,PTE\,content\,in\,soil} \tag{4}$$

$$BCF = \frac{PTE\,content\,in\,root}{Total\,PTE\,content\,in\,soil} \tag{5}$$

4.2.2 Root-to-shoot plant transfer factor

The translocation factor (TF) is a crucial parameter for assessing the phytoremediation potential of plant species. It quantifies the upward mobility of elements from belowground biomass to aerial parts, such as shoot, stem, or leaf ((Eq. (6)). A TF of 1 is recommended for categorizing a plant species with phytoremediation potential. The TF describes the characteristic nature of hyper-accumulating species and is a decisive parameter for classifying plant species for phytoremediation.

$$TF = \frac{PTE\,content\,in\,shoot}{PTE\,content\,in\,soil} \qquad (6)$$

4.3 Phytoremediation mechanisms by plant species

Phytoremediation is actually a generic term for numerous mechanisms through which plants actually use to remove contaminants from polluted environments. Plants may break down or degrade organic pollutants or remove and stabilize metals at the contaminated sites. This may be done through one or a combination of the methods described below. The methods used to phytoremediate heavy metal contaminants differ slightly from those used to remediate sites polluted with organic contaminants. Many techniques and applications are represented under phytoremediation, and each has its own mechanism of action for phytoremediating metal-polluted environment (**Figure 1**). They differ in the way plants can remove, immobilize, or degrade contaminants, as well as the type of contaminants that the plant species can target.

Phytoextraction is a process whereby plant roots absorb pollutants and then accumulate them in their aboveground tissues [117]. Phytoextraction is a frequently employed technique in remediation practices, involving the collection and disposal of contaminated plants, or, alternatively, the extraction of valuable pollutants for commercial purposes, as described by McIntyre [118]. According to the research conducted by Pathak et al. [103], it has been observed that over 500 plant species possess the ability to collect significantly elevated concentrations of heavy metals in their aboveground structures.

Phytodegradation is a process that involves the utilization of plants and their associated enzymes to break down organic contaminants, typically occurring within the plant tissues [118]. These substances are effective for chemical compounds that exhibit mobility within plants, such as herbicides [119].

Phytostabilization is a remediation technique that involves the utilization of specific plant species to mitigate the movement of contaminants, particularly heavy metals, and hinder their dispersion, such as by leaching into groundwater or entering the food chain. This is achieved by diminishing their bioavailability [117].

Phytovolatilization refers to the process through which plants ingest pollutants, such as mercury (Hg) and selenium (Se), convert them into volatile forms and subsequently release them into the environment as gases [120].

Phyto-stimulation, often referred to as rhizo-degradation, is a process wherein plants boost soil microbial activity in the root zone to facilitate the decomposition of organic pollutants, such as petroleum hydrocarbons [119]. Plants secrete sugars and acids, which serve as stimulants for microbial activity, leading to the process of biodegradation of organic pollutants.

Nevertheless, phytoremediation is subject to several restrictions, such as the extended duration it necessitates and the need for meticulous agronomic maintenance. Gerhardt et al. [121] argue that the process of phytoremediation, particularly for inorganic contaminants, can result in a substantial time burden due to the delayed reutilization of the land compared to conventional techniques. The consequence of this is the immobilization of capital, a factor that necessitates careful consideration during the process of cost analysis.

Furthermore, the efficacy of the technique is constrained to pollution levels that are classified as low to moderate. Additionally, the selection of specific plant species is necessary, taking into consideration their suitable root depth and contact area. Ultimately, the implementation of suitable agricultural techniques, encompassing crop management, harvesting, processing, and disposal, is important in order to effectively mitigate the potential re-release of toxins into the surrounding ecosystem.

5. Concluding remarks and future research

Mining has significant environmental and social impacts, including soil degradation, water quality issues, and adverse effects on fish habitats. To achieve long-term sustainability, the mining industry must restore and rehabilitate polluted sites throughout the project life cycle. Metal oxides, such as Al, Fe, and Mn, play a crucial role in the mobility of PTEs in soils due to their large surface areas. Soil organic matter influences PTE mobilization through carbonates, organometallic complexes, and other factors. Green remediation, a low-cost approach using plants, could address heavy metal contamination issues, reduce food chain contamination, limit

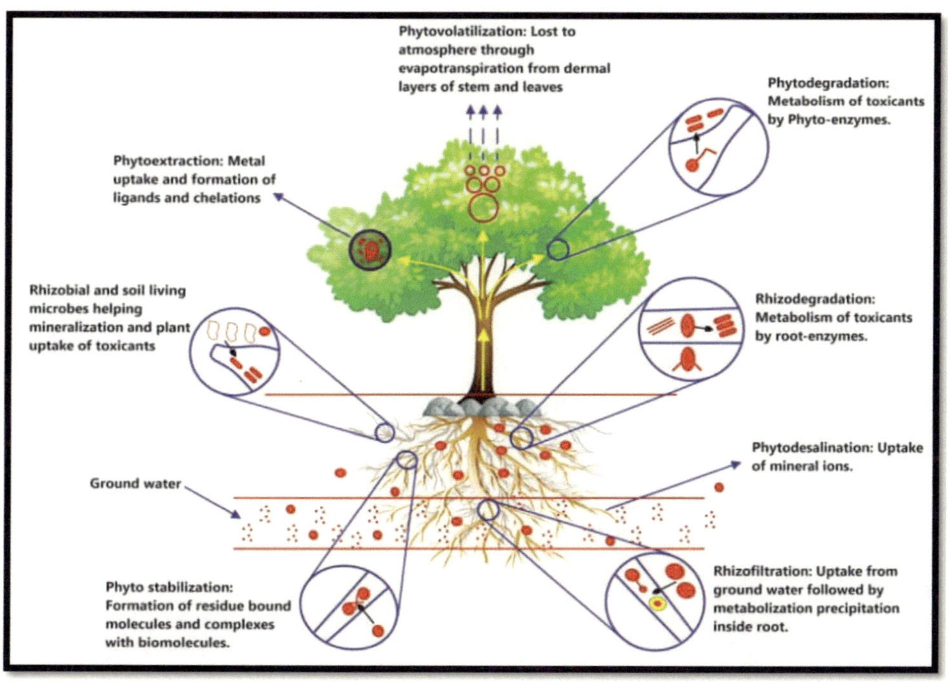

Figure 1.
An overview of different mechanisms of action for phytoremediating metal-polluted environment.

metal transfer into surface and groundwater, and guarantee a metal-contaminated-free environment. Revegetation may also aid in carbon sequestration, regulate wind and water erosion, and conserve soil fertility. Further research is needed to expand revegetation's applicability in Ghana.

Author details

Albert Kobina Mensah* and Emmanuel Amoakwah
Council for Scientific and Industrial Research-Soil Research Institute, Kumasi, Ghana

*Address all correspondence to: albert.mensah@rub.de; albertkobinamensah@gmail.com

References

[1] Mensah AK, Mahiri IO, Owusu O, Mireku OD, Wireko I, Kissi EA. Environmental impacts of mining: A study of mining communities in Ghana. Applied Ecology and Environmental Sciences. 2015;**3**(3):81-94. DOI: 10.12691/aees-3-3-3

[2] Mensah AK, Tuokuu FXD. Polluting our rivers in search of gold: How sustainable are reforms to stop informal miners from returning to mining sites in Ghana? Frontiers in Environmental Science. 2023;**11**:1154091

[3] Mensah AK, Marschner B, Shaheen SM. Biochar, compost, iron oxide, manure, and inorganic fertilizer affect bioavailability of arsenic and improve soil quality of an abandoned arsenic-contaminated gold mine spoil. Ecotoxicology and Environmental Safety. 2022;**234**(234). DOI: 10.1016/j.ecoenv.2022.113358

[4] Jeyasundar PGSAA, Ali A, Muhammad Azeem A, et al. Green remediation of toxic metals contaminated mining soil using bacterial consortium and Brassica juncea. Environmental Pollution. 2021;**277**:116789

[5] Palansooriya KN, Shaheen SM, Chen SS, Tsang DCW, Hashimoto Y, Hou D, et al. Soil amendments for immobilization of potentially toxic elements in contaminated soils: A critical review. Environment International. 2020;**134**(November 2019):105046. DOI: 10.1016/j.envint.2019.105046

[6] Hussain MM, Bibi I, Ali F, et al. The role of various ameliorants on geo-chemical arsenic distribution and CO2-carbon efflux under paddy soil conditions. Environmental Geochemistry and Health. 2022. DOI: 10.1007/s10653-021-01196-3

[7] Wuana RA, Okieimen FE. Heavy metals in contaminated soils: A review of sources, chemistry, risks and best available strategies for remediation. ISRN Ecology. 2011;**2011**:1-20. DOI: 10.5402/2011/402647

[8] Komárek M, Vaněk A, Ettler V. Chemical stabilization of metals and arsenic in contaminated soils using oxides - a review. Environmental Pollution. 2013;**172**:9-22. DOI: 10.1016/j.envpol.2012.07.045

[9] Wang L, Rinklebe J, Tack FMG, Hou D. A review of green remediation strategies for heavy metal contaminated soil. Soil Use and Management. 2021;**37**:1-28. DOI: 10.1111/sum.12717

[10] Hussain MM, Bibi I, Niazi NK, Nawaz MF, Rinklebe J. Impact of organic and inorganic amendments on arsenic accumulation by rice genotypes under paddy soil conditions: A pilot-scale investigation to assess health risk. Journal of Hazardous Materials. 2021;**420**(June):126620. DOI: 10.1016/j.jhazmat.2021.126620

[11] El-Naggar A, Shaheen SM, Hseu ZY, Wang SL, Ok YS, Rinklebe J. Release dynamics of As, Co, and Mo in a biochar treated soil under pre-definite redox conditions. Science of the Total Environment. 2019;**657**:686-695. DOI: 10.1016/j.scitotenv.2018.12.026

[12] Lemonte JJ, Stuckey JW, Sanchez JZ, Tappero R, Rinklebe J, Sparks DL. Sea level rise induced arsenic release from historically contaminated coastal soils. Environmental Science and Technology. 2017;**51**(11):5913-5922. DOI: 10.1021/acs.est.6b06152

[13] Guo L, Warnken KW, Santschi PH. Retention behavior of dissolved uranium during ultrafiltration: Implications for colloidal U in surface waters. Marine Chemistry. 2007;**107**(2):156-166. DOI: 10.1016/j.marchem.2007.06.017

[14] Yang X, Hinzmann M, Pan H, Wang J, Bolan N, Daniel C, et al. Pig carcass-derived biochar caused contradictory effects on arsenic mobilization in a contaminated paddy soil under fluctuating controlled redox conditions. Journal of Hazardous Materials. 2022;**421**(July 2021):126647. DOI: 10.1016/j.jhazmat.2021.126647

[15] Nwaichi EO, Wegwu MO, Onyeike EN. Phytoextracting cadmium and copper using Mucuna pruriens. African Journal of Plant Science. 2009;**3**(12):277-282

[16] Puga AP, Abreu CA, Melo LCA, Beesley L. Biochar application to a contaminated soil reduces the availability and plant uptake of zinc, lead and cadmium. Journal of Environmental Management. 2015;**159**:86-93. DOI: 10.1016/j.jenvman.2015.05.036

[17] Antoniadis V, Levizou E, Shaheen SM, Ok YS, Sebastian A, Baum C, et al. Trace elements in the soil-plant interface: Phytoavailability, translocation, and phytoremediation–a review. Earth-Science Reviews. 2017;**171**(October 2016):621-645. DOI: 10.1016/j.earscirev.2017.06.005

[18] Bissen M, Frimmel FH, Ag C. Arsenic – A review. Part I: Occurrence, toxicity, speciation, mobility. Clean. 2003;**31**:9-18

[19] Violante A, Cozzolino V, Perelomov L, Caporale AG, Pigna M. Mobility and bioavailability of heavy metals and metalloids in soil environments. Journal of Soil Science and Plant Nutrition. 2010;**10**(3):268-292. DOI: 10.4067/S0718-95162010000100005

[20] Chen Y, Jiang X, Xiao K, Shen N, Zeng RJ, Zhou Y. Enhanced volatile fatty acids (VFAs) production in a thermophilic fermenter with stepwise pH in-crease–investigation on dissolved organic matter transformation and microbial commu-nity shift. Water Research. 2017;**112**:261-268

[21] Mamindy-Pajany Y, Hurel C, Marmier N, Roméo M. Arsenic (V) adsorp-tion from aqueous solution onto goethite, hematite, magnetite and zero-valent iron: Ef-fects of pH, concentration and reversibility. Desalination. 2011;**281**(1):93-99. DOI: 10.1016/j.desal.2011.07.046

[22] Beiyuan J, Awad YM, Beckers F, Tsang DCW, Ok YS, Rinklebe J. Mobility and phytoavailability of As and Pb in a contaminated soil using pine sawdust biochar under systematic change of redox conditions. Chemosphere. 2017;**178**:110-118. DOI: 10.1016/j.chemosphere.2017.03.022

[23] Al-Abed SR, Jegadeesan G, Purandare J, Allen D. Arsenic release from iron rich mineral processing waste: Influence of pH and redox potential. Chemosphere. 2007;**66**(4):775-782

[24] Catalano JG, Park C, Fenter P, Zhang Z. Simultaneous inner- and outer-sphere arsenate sorption on corundum and hematite. Geochimica et Cosmochimica Acta. 2008;**72**(8):1986-2004

[25] Martin M, Celi L, Barberis E, Violante A, Kozac LM, Huang PM. Effect of humic acid coating on arsenic adsorption on ferrihydrite-kaolinite mixed systems. Canadian Journal of Soil Science. 2009;**89**:421-434 Martiñá-Prieto

[26] Violante A, Krishnamurti GSR, Pigna M. Mobility of trace elements in soil environments. In: Violante A, Huang PM, Gadd GM, editors. Biophysico-Chemical Processes of Metals and Metalloids in Soil Environments. Hoboken, NJ: John Wiley & Sons; 2008. pp. 169-213

[27] O'Reilly SE, Strawn DG, Sparks DL. Residence time effects on arsenate adsorption/desorption mechanisms on goethite. Soil Science Society of America Journal. 2001;**65**:67-77

[28] Du Laing G, Rinklebe J, Vandecasteele B, Meers E, Tack FMG. Trace metal behaviour in estuarine and riverine floodplain soils and sediments: A review. Science of the Total Environment. 2009;**407**(13):3972-3985. DOI: 10.1016/j.scitotenv.2008.07.025

[29] Shaheen SM, Rinklebe J, Frohne T, White JR, DeLaune RD. Biogeochemical factors governing cobalt, nickel, selenium, and vanadium dynamics in periodically flooded Egyptian North Nile delta rice soils. Soil Science Society of America Journal. 2014;**78**(3):1065-1078. DOI: 10.2136/sssaj2013.10.0441

[30] Rinklebe J, Shaheen SM, Yu K. Release of As, Ba, Cd, Cu, Pb, and Sr under pre-definite redox conditions in different rice paddy soils originating from the USA and Asia. Geoderma. 2016;**270**:21-32

[31] Niazi NK, Bibi I, Shahid M, Ok YS, Burton ED, Wang H, et al. Arsenic removal by perilla leaf biochar in aqueous solutions and groundwater: An integrated spectroscopic and microscopic examination. Environmental Pollution. 2018;**232**:31-41. DOI: 10.1016/j.envpol.2017.09.051

[32] Mensah AK, Bernd Marschner KB, Eric Stemn SMS, Rinklebe J. Arsenic in gold mining wastes: An environmental and human health threat in Ghana. In: Global Arsenic Hazards. 1st ed. Springer Nature; 2023. pp. 0-13. DOI: 10.1007/978-3-642-03503-6

[33] Dybowska A, Farago M, Valsami-Jones E, Thornton I. Operationally defined associations of arsenic and copper from soil and mine waste in south-West England. Chemical Speciation and Bioavailability. 2005;**17**(4):147-160. DOI: 10.3184/095422906783438811

[34] Wenzel WW, Kirchbaumer N, Prohaska T, Stingeder G, Lombi E, Adriano DC. Arsenic fractionation in soils using an improved sequential extraction procedure. Analytica Chimica Acta. 2001;**436**(2):309-323. DOI: 10.1016/S0003-2670(01)00924-2

[35] Ye F, Huang X, Zhang D, Tian L, Zeng Y. Distribution of heavy metals in sediments of the Pearl River estuary, southern China: Implications for sources and historical changes. Journal of Environmental Sciences. 2012;**24**(4):579-588

[36] Zimmer D, Kiersch K, Baum C, Meissner R, Müller R, Jandl G, et al. Scale-dependent variability of As and heavy metals in a river Elbe floodplain. CLEAN - Soil, Air, Water. 2011;**39**(4):328-337. DOI: 10.1002/clen.201000295

[37] Rinklebe J, Du Laing G, Selim HM. Factors controlling the dynamics of trace metals in frequently flooded soils. In: Dynamics and Bioavailability of Heavy Metals in the Root Zone. 2011. pp. 245-270

[38] Shaheen SM, Ali RA, Abowaly ME, Rabie AEMA, El Abbasy NE, Rinklebe J. Assessing the mobilization of As, Cr, Mo, and Se in Egyptian lacustrine and

calcareous soils using sequential extraction and biogeochemical microcosm techniques. Journal of Geochemical Exploration. 2018;**191**(May):28-42. DOI: 10.1016/j.gexplo.2018.05.003

[39] Bolan N, Kunhikrishnan A, Thangarajan R, Kumpiene J, Park J, Makino T, et al. Remediation of heavy metal(loid)s contaminated soils - to mobilize or to immobilize? Journal of Hazardous Materials. 2014;**266**:141-166. DOI: 10.1016/j.jhazmat.2013.12.018

[40] Koo N, Lee SH, Kim JG. Arsenic mobility in the amended mine tailings and its impact on soil enzyme activity. Environmental Geochemistry and Health. 2012;**34**(3):337-348

[41] Hua J, Zhang C, Yin Y, Chen R, Wang X. Phytoremediation potential of three aquatic macrophytes in manganese contaminated water. Water and Environment Journal. 2012;**26**:335-342

[42] Mench M. Physico-chemical aspects and efficiency of trace element immobilization by soil amendments. In: Metal-Contaminated Soils: In Situ Inactivation and Phytores-Toration. 1998. pp. 151-182

[43] Hartley W, Lepp NW. Remediation of arsenic contaminated soils by iron-oxide application, evaluated in terms of plant productivity, arsenic and phytotoxic metal uptake. Science of the Total Environment. 2008;**390**(1):35-44. DOI: 10.1016/j.scitotenv.2007.09.021

[44] Kunhikrishnan A, Choppala G, Seshadri B, Wijesekara H, Bolan NS, Mbene K, et al. Impact of wastewater derived dissolved organic carbon on reduction, mobility, and bioavailability of As (V) and Cr (VI) in contaminated soils. Journal of Environmental Management. 2017;**186**:183-191

[45] Prasad MNV. Plants that accumulate and/or exclude toxic trace elements play an important role in phytoremediation. In: Prasad MNV, Sajwan KS, Naidu R, editors. Trace Elements in the Environment: Biogeochemistry, Biotechnology and Bioremediation. Boca Raton, FL: CRC Press; 2006. pp. 523-547

[46] Mensah AK, Marschner B, Shaheen SM, Wang J, Wang SL, Rinklebe J. Arsenic contamination in abandoned and active gold mine spoils in Ghana: Geochemical fractionation, speciation, and assessment of the potential human health risk. Environmental Pollution. 2020;**261**:114116. DOI: 10.1016/j.envpol.2020.114116

[47] Hu L, Zhang Z, Xiang Z, Yang Z. Exogenous application of citric acid ameliorates the adverse effect of heat stress in tall fescue (Lolium arundinaceum). Frontiers in Plant Science. 2016;**7**:179

[48] Shaheen SM, Rinklebe J. Phytoextraction of potentially toxic elements by Indian mustard, rapeseed, and sunflower from a contaminated riparian soil. Environmental Geochemistry and Health. 2015;**37**(6):953-967. DOI: 10.1007/s10653-015-9718-8

[49] Maejima Y, Nagatsuka S, Higashi T. Application of the crystallinity ratio of free iron oxides for dating soils developed on the raised coral reef terraces of Kikai and Minami-Daito Islands, Southwest Japan. The Quaternary Research (Daiyonki-Kenkyu). 2002;**41**(6):485-493

[50] DeSisto SL, Jamieson HE, Parsons MB. Subsurface variations in arsenic mineralogy and geochemistry following long-term weathering of gold mine tailings. Applied Geochemistry. 2016;**73**:81-97. DOI: 10.1016/j.apgeochem.2016.07.013

[51] Murciego A, Pellitero Pascual E, Rodríguez González M, ÁlvarezAyuso E, García Sánchez A, Rubio Alonso F, et al. Secondary Products of Arsenopyrite in the Terrubias Mining Area. Spain: Salamanca; 2010

[52] Drewniak L, Sklodowska A. Arsenic-transforming microbes and their role in biomining processes. Environmental Science and Pollution Research. 2013;**20**(11):7728-7739. DOI: 10.1007/s11356-012- 1449-0

[53] von der Heyden BP, Benoit J, Fernandez V, Roychoudhury AN. Synchrotron X-ray radiation and the African earth sciences: A critical review. Journal of African Earth Sciences. 2020:104012. DOI: 10.1016/j.jafrearsci.2020.104012

[54] Hou D, O'Connor D, Igalavithana AD, Alessi DS, Luo J, Tsang DCW, et al. Metal contamination and bioremediation of agricultural soils for food safety and sustainability. Nature Reviews Earth & Environment. 2020;**1**(7):366-381. DOI: 10.1038/s43017-020-0061-y

[55] Redman AD, Macalady DL, Ahmann D. Natural organic matter affects arsenic speciation and sorption onto hematite. Environmental Science & Technology. 2002;**36**:2889-e2896

[56] Shaheen SM, Shams MS, Khalifa MR, El-Dali MA, Rinklebe J. Various soil amendments and environmental wastes affect the (im)mobilization and phytoavailability of potentially toxic elements in a sewage effluent irrigated sandy soil. Ecotoxicology and Environmental Safety. 2017;**142**(April):375-387. DOI: 10.1016/j.ecoenv.2017.04.026

[57] McGowen SL, Basta NT, Brown GO. Use of diammonium phosphate to reduce heavy metal solubility and transport in smelter-contaminated soil. Journal of Environmental Quality. 2001;**30**:493-500

[58] Pueyo M, López-Sanchez JF, Rauret G. Assessment of CaCl2, NaNO3 and NH4NO3 extraction procedures for the study of Cd, Cu, Pb and Zn extractability in contaminated soils. Analytica Chimica Acta. 2004;**504**:217-226

[59] McLaughlin MJ, Tiller KG, Beech TA, Smart MK. Soil salinity causes elevated cadmium concentrations in field-grown potato tubers. Journal of Environmental Quality. 1994;**23**:1013-1018

[60] Holland JE, Bennett AE, Newton AC, White PJ, McKenzie BM, George TS, et al. Liming impacts on soils, crops and biodiversity in the UK: A review. Science of the Total Environment. 2018;**610**:316-332

[61] Lebrun M, Van Poucke R, Miard F, Scippa GS, Bourgerie S, Morabito D, et al. Effects of carbon-based materials and redmuds on metal(loid) immobilization and growth of Salix dasyclados Wimm. On a former mine technosol contaminated by arsenic and lead. Land Degradation and Development. 2020;**32**:467-481. DOI: 10.1002/ldr.3726

[62] Zou Q, An W, Wu C, Li W, Fu A, Xiao R, et al. Red mud-modified biochar reduces soil arsenic availability and changes bacterial composition. Environmental Chemistry Letters. 2018;**16**(2):615-622. DOI: 10.1007/s10311-017-0688-1

[63] Mensah AK, Marschner B, Antoniadis V, Stemn E, Shaheen SM, Rinklebe J. Human health risk via soil ingestion of potentially toxic elements and remediation potential of native plants near an abandoned mine spoil in Ghana. Science of the Total Environment. 2021;**798**. DOI: 10.1016/j.scitotenv.2021.149272

[64] Sheoran V. Soil reclamation of abandoned mine land by revegetation: A review. International Journal of Soil, Sediment and Water. 2010;**3**(2):13

[65] Reeves RD, Baker AJM, Jaffré T, Erskine PD, Echevarria G, van der Ent A. A global database for plants that hyperaccumulate metal and metalloid trace el-ements. The New Phytologist. 2018;**218**(2):407-411. DOI: 10.1111/nph.14907

[66] Jiang J, Yuan M, Xu R, Bish DL. Mobilization of phosphate in variable-charge soils amended with biochars derived from crop straws. Soil and Tillage Research. 2015;**146**:139-147. DOI: 10.1016/j.still.2014.10.009

[67] Antoniadis V, Shaheen SM, Stärk HJ, Wennrich R, Levizou E, Merbach I, et al. Phytoremediation potential of twelve wild plant species for toxic elements in a contaminated soil. Environmental International. 2021;**146**:106233. DOI: 10.1016/j.envint.2020.106233

[68] Mensah AK. Role of revegetation in restoring fertility of degraded mined soils in Ghana: A review. International Journal of Biodiversity and Conservation. 2015;**7**(2):57-80. DOI: 10.5897/ijbc2014.0775

[69] Mahar A, Wang P, Ali A, Awasthi MK, Lahori AH, Wang Q, et al. Challenges and opportunities in the phytoremediation of heavy metals contami-nated soils: A review. Ecotoxicology and Environmental Safety. 2016;**126**:111-121

[70] Baker AJM, McGrath SP, Reeves RD, Smith JAC. Metal hyperaccumulator plants: A review of the ecology and physiology of a biological resource for phytoremediation of metal polluted soils. In: Terry N, Banuelos GS, editors. Phytoremediation of Contaminants in Soil and Water. Boca Raton (FL): CRC Press; 2000. pp. 85-107

[71] Blum WEH. Problems of soil conservation. Nature and environment series 39, Council of Europe, Strasbourg. In: Towards Sustainable Land Use: Furthering Cooperation between People and Institutions. 1(9). Selected Papers of the Conference of ISCO. Reiskirchen, Germany: Catena Verlag; Adv. Geoecol. (31): 755-757

[72] Sheoran V, Sheoran AS, Poonia P. Soil reclamation of abandoned mine land by revegetation: A review. International Journal of Soil, Sediment and Water. 2010;**3**:1-21

[73] Bradshaw AD. Restoration of mined lands using natural processes. Ecological Engineering. 1997;**8**:255-269

[74] Kobayashi S. Landscape Rehabilitation of Degraded Tropical Forest Ecosystems. A Case Study of the CIFOR/Japan Project in Indonesia and Peru2004. pp. 1-2

[75] Grigg AH, Sheridan GJ, Pearce AB, Mulligan DR. The effect of organic mulch amendments on the physical and chemical properties and revegetation success of a sa-line-sodic mine spoil from Central Queensland. Australia. 2006:1-9

[76] Bewley JD, Black N. Physiology and Biochemistry of Seeds in Relation to Germination. Berlin: Springer-Verlag; 1978. p. 2

[77] Harwood MR. The selection of new pasture grasses for the revegetation of open cut coal mines in central Queensland [PhD thesis]. Department of Agriculture, University of Australia; 1998

[78] Coppin P, Hermy M, Honnay O. Impact of habitat quality of Forest Plant

species colonization. Forest Ecology and Management. 2000;**115**:157-170

[79] Owen OS, Chiras DD, Reganold. Natural Resource Conservation for Sustainable Future. 7th ed. 1998. pp. 503-504

[80] Wong MH. Ecological restoration of mine degraded soils, with emphasis on met-al contaminated soils. Chemosphere. 2003;**50**:775-780

[81] Blaylock MJ, Salt DE, Dushenkov S, Zakharova O, Gussman C, Kapulnik Y, et al. Enhanced accumulation of Pb in Indian mustard by soil-applied chelating agents. Environmental Science & Technology. 1997;**35**(9):184A-185A

[82] Ensley BD. Rationale for use of phytoremediation. In: Raskin I, Ensley BD, editors. Phytoremediation of Toxic Metals: Using Plants to Clean up the Environment. John Wiley and Sons: New York; 2000. pp. 3-11

[83] Huang JW, Chen J, Berta WR, Cunningham SD. Phytoremediation of lead contaminated soil: Role of synthetic chelates in lead phytoextraction. Environmental Science & Technology. 1997;**31**(3):800-805

[84] Van der Lelie D, Schwitduebel J, Glass DJ, Vangronsveld J, Baker A. Assessing phytoremediation progress in the United States and Europe. Environmental Science & Technology. 2001;**35**(21):447A-452A

[85] Watanabe ME. Phytoremediation on the brink of commercialization. Environmental Science & Technology. 1997;**31**:182-186

[86] Tordoff GM, Baker AJM, Willis AJ. Current approaches to the revegetation and reclamation of metalliferous mine wastes. Chemosphere. 2000;**41**(1-2):219-228. DOI: 10.1016/S0045-6535(99)00414-2

[87] Conesa HM, Faz Á, Arnaldos R. Initial studies for the phytostabilization of a mine tailing from the Cartagena-La Union Mining District (SE Spain). Chemosphere. 2007;**66**(1):38-44. DOI: 10.1016/j.chemosphere.2006.05.041

[88] Li MS. Ecological restoration of mine land with particular reference to the metalliferous mine wasteland in China: A review of research and practice. The Science of the Total Environment. 2006;**357**:38-53

[89] Mendez MO, Maier RM. Phytoremediation of mine tailings in temperate and arid environments. Reviews in Environmental Science and Biotechnology. 2008a;**7**:47-59

[90] Mendez MO, Maier RM. Phytostabilization of mine tailings in arid and semi-arid environments- an emerging remediation technology. Environmental Health Perspectives. 2008b;**116**(3):278-283

[91] Ramírez O, de la Campa AMS, Sánchez-Rodas D, Jesús D. Hazardous trace elements in thoracic fraction ofairborne particulate matter: Assessment oftemporal variations, sources, and health risks in a megacity. Science of the Total Environment. 2020;**710**:136344

[92] Shaheen SM, Antoniadis V, Kwon E, Song H, Wang SL, Hseu ZY, et al. Soil contamination by potentially toxic elements and the associated human health risk in geo- and anthropogenic contaminated soils: A case study from the temperate region (Germany) and the arid region (Egypt). Environmental Pollution. 2020;**262**:114312. DOI: 10.1016/j.envpol.2020.114312

[93] Bansah KJ, Dumakor-Dupey NK, Stemn E, Galecki G. Mutualism,

commensalism or parasitism? Perspectives on tailings trade between large-scale and artisanal and small-scale gold mining in Ghana. Resources Policy. 2018;**57**:246-254. DOI: 10.1016/j.resourpol.2018.03.010

[94] Armah FA, Gyeabour EK. Health risks to children and adults residing in riverine environments where surficial sediments contain metals generated by active gold-mining in Ghana. Toxicology Research. 2013;**29**(1):69-79. DOI: 10.5487/TR.2013.29.1.069

[95] Narh CT, Dzamalala CP, Mmbaga BT, Menya D, Mlombe Y, Finch P, et al. Geophagia and risk of squamous cell esophageal cancer in the African esophageal cancer corridor-findings from the ESCCAPE multi-country case-control studies. International Journal of Cancer. 2021;**149**:1274-1283

[96] Bundschuh J, Schneider J, Alam MA, Niazi NK, Herath I, Parvez F, et al. Seven potential sources of arsenic pollution in Latin America and their environmental and health impacts. Science of the Total Environment. 2021;**780**(May):146274. DOI: 10.1016/j.scitotenv.2021.146274

[97] Bortey-Sam N, Nakayama SMM, Ikenaka Y, Akoto O, Baidoo E, Yohannes YB, et al. Human health risks from metals and metalloid via consumption of food animals near gold mines in Tarkwa, Ghana: Estimation of the daily intakes and target hazard quotients (THQs). Ecotoxicology and Environmental Safety. 2015;**111**:160-167. DOI: 10.1016/j.ecoenv.2014.09.008

[98] Antoniadis V, Golia EE, Liu YT, Wang SL, Shaheen SM, Rinklebe J. Soil and maize contamination by trace elements and associated health risk assessment in the industrial area of Volos, Greece. Environment International. 2019;**124**:79-88

[99] Mulligan C, Bronstein JM. Wilson disease: An overview and approach to management. Neurologic Clinics. 2020;**38**(2):417-432. DOI: 10.1016/j.ncl.2020.01.005

[100] Apostoli P, Catalani S. Metal Ions Affecting Reproduction and Development Metal Ions in Toxicology: Effects, Interactions, Interdependencies. Vol. 11. De Gruyter; 2015. pp. 263-304

[101] Bempah CK, Ewusi A, Obiri-Yeboah S, Asabere SB, Mensah F, Boateng J, et al. Distribution of arsenic and heavy metals from mine tailings dams at Obuasi municipality of Ghana. American Journal of Engineering Research. 2013;**2**(5):61-70

[102] Bansah KJ, Addo WK. Phytoremediation potential of plants grown on reclaimed spoil lands. Ghana Mining Journal. 2016;**16**(1):68. DOI: 10.4314/gmj.v16i1.8

[103] Pathak L, Shah K. Phytoremediation of abandoned mining areas for land restoration: Approaches and technology. In: Phytorestoration of Abandoned Mining and Oil Drilling Sites. Elsevier; 2021. pp. 33-56

[104] Petelka J, Abraham J, Bockreis A, Deikumah JP, Zerbe S. Soil heavy metal (loid) pollution and phytoremediation potential of native plants on a former gold mine in Ghana. Water, Air, and Soil Pollution. 2019;**230**(11):1-16

[105] Sheoran V, Choudhary RP. Phytostabilization of mine tailings. In: Phytorestoration of Abandoned Mining and Oil Drilling Sites. Elsevier; 2021. pp. 307-324

[106] Ashraf MA, Maah MJ, Yusoff I. Evaluation of natural phytoremediation process occurring at ex tin mining catchment. Chiang Mai Journal of Science. 2013;**40**(2):198-213

[107] Liu S, Ali S, Yang R, Tao J, Ren B. A newly discovered cd-hyperaccumulator Lan- tana camara L. Journal of Hazardous Materials. 2019;**371**:233-242

[108] Issaka S, Ashraf MA. Phyto restoration of mine spoiled: Evaluation of natural phytoremediation process occurring at ex-tin mining catchment. In: Phytorestoration of Abandoned Mining and Oil Drilling Sites. Elsevier; 2021. pp. 219-248

[109] Williams HT, Bellito MW. Revegetation of historic high altitude mining wastes in the western United States. In: Land Reclamation: Achieving Sustainable Benefits. Fox, Moore, Mclntosh, editors. 1998. pp. 193-194

[110] Addai P, Mensah AK, Sekyi-Annan E, Owusu EA. Biochar, compost and / or NPK fertilizer affect the uptake of potentially toxic elements and promote the yield of lettuce grown in an abandoned gold mine tailing. Journal of Trace Elements and Minerals. 2023;**4**:100066

[111] Baker AJM. Accumulators and excluders-strategies in the response of plants to heavy metals. Journal of Plant Nutrition. 1981;**3**:643-654

[112] Blaylock MJ, Huang JW. Phytoextraction of metals. In: Raskin I, Ensley BD, editors. Phytoremediation of Toxic Metals: Using Plants to Clean Up the Environment. New York: Wiley; 2000. pp. 53-70

[113] Cunningham SD, Ow DW. Promises and prospects of phytoremediation. Plant Physiology. 1996;**110**:715-719

[114] Lasat MM, Baker AJM, Kochian LV. Altered Zn compartmentation in the root symplasm and stimulated Zn absorption into the leaf as mechanisms involved in Zn hyper accumulation in Thlaspi caerulescens. Plant Physiology. 1998;**118**:875-883

[115] Fitz WJ, Wenzel WW. Arsenic transformations in the soil-rhizosphere-plant system: Fundamentals and potential application to phytoremediation. Journal of Biotechnology. 2002;**99**(3):259-278. DOI: 10.1016/S0168-1656(02)00218-3

[116] Yoon J, Cao X, Zhou Q, Ma LQ. Accumulation of pb, cu, and zn in native plants growing on a contaminated Florida site. Science of the Total Environment. 2006;**368**:456-464

[117] Sharma P, Pandey S. Status of phytoremediation in world scenario. International Journal of Environmental Bioremediation & Biodegradation. 2014;**2**(4):178-191

[118] McIntyre T. Phytoremediation of heavy metals from soils. Advances in Biochemical Engineering/Biotechnology. 2003;**78**:97-123

[119] Pilon-Smits E. Phytoremediation. Annual Review of Plant Biology. 2005;**56**:15-39

[120] Ali A, Guo D, Jeyasundar PGSA, Li Y, Xiao R, Du J, et al. Application of wood biochar in polluted soils stabilized the toxic metals and enhanced wheat (Triticum aestivum) growth and soil enzymatic activity. Ecotoxicology and Environmental Safety. 2019;**184**:109635

[121] Gerhardt KE, Gerwing PD, Greenberg BM. Opinion: Taking phytoremediation from proven technology to accepted practice. Plant Science. 2017;**256**:170-185

Chapter 4

Territorial Governance and Social Participation for the Remediation of Contaminated Soils

José G. Chan-Quijano, Karla L. Torres-López and Ricardo Rodríguez-Cabrera

Abstract

Territorial governance evaluates the management and government of public policies from a perspective of social participation for the dynamics of territorial cohesion of contaminated soils for their initiation, implementation, and evaluation, which is why, when applying governance, it monitors the impact of socio-environmental conflicts generated by soil contamination, since each area and region is different and, therefore, different management alternatives and guidelines must be applied for the remediation of areas affected by contaminants, such as oil spills, pesticides, and heavy metals. On the other hand, residents must be considered and involved in carrying out the biorecovery and stabilization of contaminated soils.

Keywords: management, sustainability, conflicts, pollution, socio-environmental, public policies

1. Introduction

Soil is an essential, nonrenewable resource, providing food and supporting aquatic systems, vegetation, and all biotic and nonabiotic components. The importance of soil is increasing because this natural resource is threatened by pollution, which has significantly reduced its quality [1, 2].

Soil contamination is sometimes due to natural processes (volcanic eruptions, earthquakes, and tsunamis, to name a few) and due to anthropogenic activities, such as extraction of metals (trace and heavy metals), chemical products (pesticides, oil, polymers, plastics, and other waste), and radioactive waste (nuclear energy generation and byproducts from nuclear technology) [3].

These substances are carcinogenic, teratogenic, and mutagenic. When they have high concentrations for long periods of time, they become potentially toxic to socio-ecosystems because they accumulate and get biomagnified until they cause damage to nature and humans [4, 5]. In this sense, the effects of pollution on the environment and humans can be slow (chronic) or can act quickly (acute) and, therefore, multiple studies are required over time to know the effects and simultaneous adverse reactions on environmental and human health [6].

 IntechOpen

Likewise, the effects of pollution occur at all spatial scales, and it must be considered that the environment is continuous and does not obey political limits [6–8]. For many governments, implementing public environmental policies to remediate contaminated soils is becoming a priority because, within the framework of planetary limits [9], six of the nine limits are transgressed, suggesting that the planet Earth is no longer a safe operating space for humanity [10].

Given this concern, alternatives that are friendly to the natural environment are sought to achieve, in this case, the remediation of contaminated soils, for example, there are biotechnologies such as green and sustainable remediation where the processes are considered biological-ecological, social-anthropological, and economic-administrative for remediation and restoration projects in contaminated areas (**Figure 1**) [11–14].

Likewise, for remediation and restoration processes, people must be considered and involved to create scenarios where cooperation, communication, and trust between actors (society, business, government, academia, and civil associations) are visualized to create an ideal environment for solving problems, conflicts, and socio-environmental repercussions due to soil contamination and, thereby, reducing and solving the socio-environmental crisis of the territory [15–17].

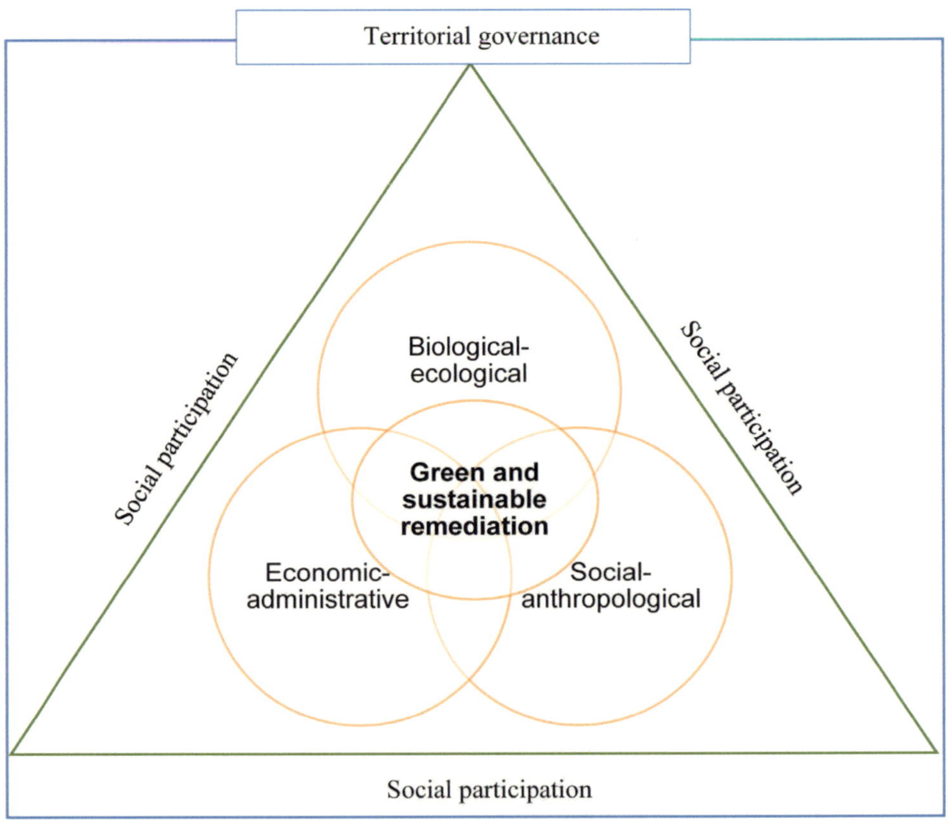

Figure 1.
Linking processes between social participation, the axes of sustainable remediation, and territorial governance.

Although there is technical and scientific evidence on soil contamination worldwide [18], it seems that this does not matter and that the real concern is to continue generating more viable economic development at the expense of nature, that is, underestimating the biocapacity and importance of recovering the affected soils [14, 19].

Preventing the environmental, social, and economic collapse of a territory due to soil contamination must be focused on the multiplicity of biotechnologies that can be implemented for the remediation of the affected areas, taking into account the different spatial, temporal, and social scales within of the complex framework of adaptive (sub)system and socio-ecological systems so that, with this, territorial governance can be managed [20, 21].

Territorial governance refers to the complex, multifaceted, and plural process between the different actors for decision-making, that is, it seeks adaptive efficiency and requires flexibility, experimentation, and learning with different interests and the common good within a given space [22, 23] in this case, areas impacted by pollution. Whereas the authorities seek to apply remediation strategies.

Now, territorial governance is required not only to consider political management, but the coordination of one or more modes of governance must be considered using different instruments, methods, and strategies under social participation to overcome governance failures and, at the same time, to achieve sustainable remediation of contaminated soils [24–26]. Therefore, the objective of this work is to analyze territorial governance as a strategy in the search for social participation in the remediation of contaminated soils.

2. Territorial governance, social participation, and contaminated soils

Governance seeks cooperation between governments, public and private administrations, nongovernmental actors, and civil society in the development of public policies for a common good [27]. Furthermore, by applying good governance, the management objectives will be effective and successful, considering international agreements, to use processes, techniques, and methods to provide a solution to a given problem [28].

In this sense, social participation supports decision-making and solutions at the local, regional, and even territorial level in a participatory manner, since the effective participation of society manages to contribute to legal instruments and territorial planning plans, that is, if citizens are considered, their contributions to caring for the environment and the economy of the territory would be favorable [29, 30].

Likewise, this social participation collaborates, empowers, and creates social management within public environmental policies on soil contamination issues, which generates debates around decision-making processes, especially those related to physical and chemical methods, biological and soil resource management, as well as the damage caused to the natural resources of the territory (**Figure 2**) [31–34].

When applying territorial governance, the participation of public administration, political science, and social participation is considered within the strategies to shape the behavior of the actors to generate strategies and rules for the application of biotechnologies in the remeasurement process of contaminated soils and the environmental regulation of companies is not compromised toward the sanitation of the affected socio-ecosystem [8, 25].

Figure 2.
Social participation of a community in Tabasco, Mexico, regarding its natural resources in the territory. Photographs: José G. Chan-Quijano.

However, it must be considered that soil contamination may be different in the concentration levels of the contaminant, in its distribution over the territory and that the contaminant will be different, for example, heavy metals, metalloids, hydrocarbons, pesticides, plastics, and emerging contaminants, to name a few. Furthermore, the biogeochemical behavior of each pollutant in the system will cause damage at different scales to natural resources and humans [31, 34–36].

Therefore, territorial governance and social participation are two conceptual and practical methods that develop strategies within remediation biotechnologies to achieve effectiveness in the recovery of contaminated soils, that is, from a political and social sphere, they can generate methods for choosing, controlling, and replacing viable techniques for the recovery of the contaminated area [37, 38].

In this context, and at the same time, a capacity for monitoring and principles of public policy will be generated in the administration of resources and environmental policies (regulatory framework and government effectiveness toward the problem), and, above all, the consideration of citizens in the participation of the remeasurement of the contaminated environment in conjunction with territorial governance and emotions toward nature (**Figure 3**).

Therefore, considering the previous principles and following up on the problem, it is crucial to develop methods and techniques to evaluate the potential risks of human exposure to contaminants and decide the concentration thresholds in soils to protect the health of the socio-ecosystems [3, 43]. Because soil contamination continues to increase and remediation practices are few, social participation must be considered to achieve sustainable remediation in conjunction with territorial governance.

Public policy principles in the remediation of contaminated soils

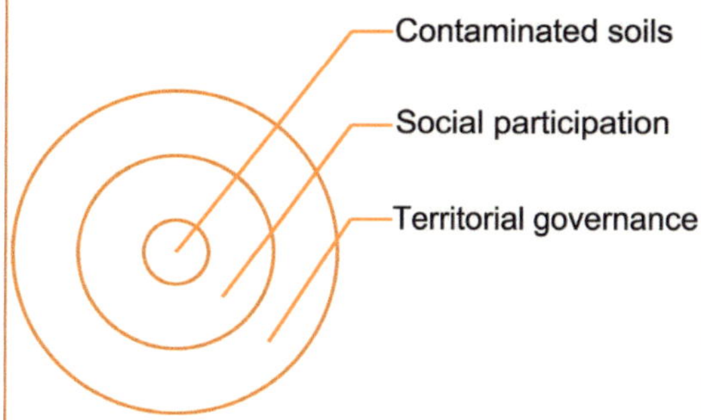

Contaminated soils

- Identification of the source of contamination.
- The diversity of soils in a single territory.
- The different contaminants and their toxicity.
- Environmental and human health.
- Use and function of the land.
- Classification of remediation biotechnology.
- Specific laws and regulations.

Social participation

- Sociology and environmental anthropology of the contaminated area.
- Dynamics and ways of life of the territory.
- Decision making in biotechnology to implement for remediation.
- Socio-environmental impact.
- Problems, conflicts, and repercussions.
- Perceptions and subjectivities of the subjects (environmental, social, and economic).
- Management and information of the area.

Territorial governance

- Socioecological justice.
- Green criminology of the area affected by pollution.
- Specific laws, rules, and regulations.
- Socio-environmental and socio-ecological responsibility.
- Political and institutional dimension in decision-making for remediation.
- Levels of government, their commitment and political will.
- State secretaries for stronger decisions on the care and remediation of contaminated environments.

Figure 3.
Principles and interaction between territorial governance, social participation, and the remediation of contaminated soils within environmental policy. Source: own elaboration based on Refs. [23, 39–42].

3. Humans and nature: the socio-environmental emergency of contaminated soils

The link between humans and nature supports knowing that epistemological and methodological forces for the metatheoretical formulation to seek the socio-environmental stability of the affected socio-ecosystems, in this case, soil contamination. Legally, this problem (company-society-government) is given attention, but with flaws in the environmental legal system since it seems that nature is absent due to monetary interest [44, 45].

When looking for alternatives for management, remediation, and restoration of contaminated soils, it is, without a doubt, taking society into account so that citizens can understand their concerns about the affected environment that surrounds them and, with this, achieve this change of thinking about its environmental rationality not only with the view that a payment will be received for compensation for the damage caused but also to achieve a democracy of the Earth, that is, that with the same social participation the socio-ecosystem affected by pollution [46–48].

According to the above, there are various biotechnologies, for example, bioremediation, phytoremediation, and electroremediation, to mention a few, but they are applied from an ecological-biological perspective for the remediation of contaminated soil. However, it is necessary to apply these techniques, but considering society and, at the same time, the territorial governance of the affected socio-ecosystem, that is, apply the ecological-biological, social-anthropological, and economic-administrative axes for sustainable remediation [11, 49, 50].

On the other hand, within territorial governance, green criminology can be brought closer to environmental policy through the studies of environmental crimes and crimes derived from soil contamination by various organic and inorganic pollutants, since these compounds are factors that threaten the natural components of the socio-ecosystems. Therefore, green criminology offers an epistemological, theoretical, and methodological framework to understand the legal framework that, in turn, will achieve successful and efficient remediation [51–53].

In this context, the environmental crisis is a civilizational crisis and human irresponsibility in the face of contamination of socio-ecological systems and is driven by a lack of being and a will to power in environmental politics, that is, there is a struggle for life [54]. Finally, we must rethink the current state of contaminated soils in the territory and, with this, achieve an alternative that goes beyond what is legal, that is, society itself is aware of the damage being caused to the ground which, many times, is irreparable.

4. Perspective of the state and social consciousness: case studies in the remediation of contaminated soils

At the international level, there are a few initiatives on the soil protection regime, including the issue of contamination and remediation measures, which could generate a serious risk to environmental and human health, as well as put at risk the long-term sustainable development of the land's soils [55–57]. In this sense, a holistic perspective must be applied where the environmental, social, and economic benefits of remediation are maximized and, with this, territorial governance and socio-environmental management of contaminated sites must be put into practice [58].

Territorial governance and socio-environmental management propose that the remediation of contaminated soils within public environmental policy should be a priority for policymakers, professional organizations, and all parties interested in environmental protection; therefore, a multidisciplinary scenario must be applied where side effects are considered, as well as socioeconomic aspects [59]. Likewise, human environmental ethics must be part of the reciprocal relationships between nature and social conscience.

Within environmental awareness, emerging collaborative contributions must be applied where the capacity for self-design of contaminated soils is based on systemic approaches that help to remediate, restore, and conserve socio-ecosystems (**Table 1**) [60, 64]. Furthermore, in the social consciousness of the subjects what is sought is the redefinition of sustainability and new rationalities with an orientation toward governance and participation with other actors so that the State can develop more specific public policies for the recovery and remediation of the contaminated soils [50].

Emerging collaborative contribution	Characteristics	Link to remediation
Care of the place	It considers the relationship of the subjects with their natural environment, which they appropriate physically and spiritually (worldview) where they establish their home and, therefore, take care of natural resources.	Examines the perception and concern about the contamination of the area and seeks remediation and restoration of the affected socio-ecosystem, being careful not to look at the area as a monetary resource.
Care of power	Examines how the subjects of the affected areas participate in decision-making with aspects of environmental public policy and territorial governance.	Explores the need to apply socio-ecological justice under the study of green criminology and environmental ethics in social awareness toward conservation.
Care of the common good	It explores collective action and provides information on recovery practices for the common good, that is, it seeks new strategies and methodologies based on nature or people themselves.	It recognizes that contaminated soils are integrated with the social system, which is why it seeks to co-manage sustainable biotechnologies by incorporating environmental policies to solve common problems.

Source: Prepared based on Refs. [60–63].

Table 1.
Collaborative contributions under social awareness for the self-design of strategies for the remediation of contaminated soils.

According to the collaborative contributions, the social awareness of the subjects and the State within their participation in territorial governance is that there are cases that have managed to apply some remediation method, for example, in Mexico, the national remediation program of contaminated sites with the objective of applying priority strategies and specific actions, as well as interinstitutional coordination tasks for the implementation or operation of the remediation of contaminated soils, taking into account participation and social well-being [65].

In Japan, they have the objective of applying technical development strategies to advance the application of viable technologies for final disposal areas of contaminated soils. Regarding regulatory aspects, managed recycling of removed soil and final disposal of removed soil will be applied. This soil will be converted to recycled material through quality adjustment and other processes to match the conditions of the materials to be used. In addition, they consider the social aspect of the care of human and environmental health and, with this, what they seek is trust and acceptability to increase social benefits and reduce the perception of risk [66, 67].

For its part, the United States is implementing best management practices where remediation technology can be used to reduce the environmental footprint and, in addition, focuses on using passive treatment systems and installing soil covers with an integration approach. Renewable energy is used on-site to recover natural resources through restoration and socio-environmental management plans [68].

Finally, and based on the examples, it is noted that there is international concern about contaminated soils, but there is still a lack of international regulation that supports the remediation and restoration of contaminated soils and, thereby, preserving the quality and functionality of the soil to ensure long-term environmental sustainability [19, 27, 48]. These synergistic actions, which consider both nature and society, provide a sustainable and efficient alternative for the State, together with territorial governance and social participation, to achieve green and sustainable remediation of contaminated soils.

5. Conclusions

Soil contamination is increasing, and remediation alternatives are not always viable due to the socio-environmental problems that are generated by society-business-government. Given this problem, the strategy of working with social participation is planned so that recovery strategies are generated from society itself, as well as the selection of the most viable remediation biotechnology for the affected area.

Social participation must be a priority in remediation projects for contaminated soils to generate strategies based on the same people to recover the affected socio-ecosystems. Likewise, working with territorial governance opens the opportunity to have principles and guidelines from environmental policy to remediate contaminated soils.

Sustainable remediation under its different axes can be an alternative for recovering the contaminated socio-ecosystem, but at the same time, it generates rationality and environmental awareness in society itself. Finally, at an international level, there is still a need to carry out remediation projects for contaminated soils where social participation is considered and, with this, a restoration of the recovered site is subsequently achieved.

Acknowledgements

The authors would like to thank the Center for Global Change and Sustainability (Centro de Cambio Global y la Sustentabilidad) and the Autonomous University of Guadalajara (Universidad Autónoma de Guadalajara) for their support for this research.

Conflict of interest

The authors declare no conflict of interest.

Author details

José G. Chan-Quijano[1*], Karla L. Torres-López[2] and Ricardo Rodríguez-Cabrera[2]

1 Center for Global Change and Sustainability (Centro del Cambio Global y la Sustentabilidad, A.C.), Calle Centenario del Instituto Juárez S/N Col. Reforma, Villahermosa, Tabasco, México

2 Departamento de petrolera, energía y ciencias básicas, Universidad Autónoma de Guadalajara. Prol. Paseo Usumacinta, Fracc. El Country, Villahermosa, Tabasco, México

*Address all correspondence to: jose.chan@ccgs.mx and guadachaqui@gmail.com

References

[1] Adame-Romero A. Contaminación Ambiental y Calentamiento Global. México: Editorial Trillas; 2010. 197 p

[2] Montaño-Arias NM, Navarro-Rangel MC, Patricio-López IC, Chimal-Sánchez E, Miguel-de la Cruz J. El suelo y su multifuncionalidad ¿qué ocurre ahí abajo? Ciencia ergo-sum. Revista Científica Multidisciplinaria de Prospectiva. 2018;**25**:1-12. DOI: 10.30878/ces.v25n3a9

[3] Koul B, Taak P. Soil pollution: Causes and consequences. In: Koul B, Taak P, editors. Biotechnological Strategies for Effective Remediation of Polluted Soils. 1st ed. Singapore: Springer; 2018. pp. 1-37. DOI: 10.1007/978-981-13-2420-8-1

[4] da Silva-Junior FC, Mesquita Cansanção Felipe MB, Farias de Castro DE, da Silva Araújo SC, Nóbrega Sisenando HC, Batistuzzo de Medeiros CR. A look beyond the priority: A systematic review of the genotoxic, mutagenic, and carcinogenic endpoints of non-priority PAHs. Environmental Pollution. 2021;**278**:116838. DOI: 10.1016/j.envpol.2021.116838

[5] Dasharathy S, Arjunan S, Basavaraju AM, Murugasen V, Ramachandran S, Keshav R, et al. Mutagenic, carcinogenic, and teratogenic effect of heavy metals. Evidence-Based Complementary and Alternative Medicine. 2022;**8011953**:1-11. DOI: 10.1155/2022/8011953

[6] Artiola JF, Brusseau ML. The role of environmental monitoring in pollution science. In: Brusseau ML, Pepper IL, Gerba CP, editors. Environmental and Pollution Science. 3rd ed. Cambridge, Massachusetts: Academic Press; 2019. pp. 149-162. DOI: 10.1016/B978-0-12-814719-1.00010-0

[7] Belov M. Territory, territoriality, and territorial politics as public law concepts. In: Belov M, editor. Territorial Politics and Secession. Federalism and Internal Conflicts. UK, London: Palgrave Macmillan, Cham; 2021. pp. 15-43. DOI: 10.1007/978-3-030-64402-4_2

[8] Xiao G, Shen S. To pollute or not to pollute: Political connections and corporate environmental performance. Journal of Corporate Finance. 2022;**74**:102214. DOI: 10.1016/j.jcorpfin.2022.102214

[9] Rockström J, Steffen W, Noone K, Persson A, Stuart CF III, Lambin E, et al. Planetary boundaries: Exploring the safe operating space for humanity. Ecology and Society. 2009;**14**:32

[10] Richardson K, Steffen W, Lucht W, Bendtsen J, Cornell SE, Donges JF, et al. Earth beyond six of nine planetary boundaries. Science. Advances. 2023;**9**:eadh2458. DOI: 10.1126/sciadv.adh245

[11] Tripathi V, Fraceto LF, Abhilash PC. Sustainable clean-up technologies for soils contaminated with multiple pollutants: Plant-microbe-pollutant and climate nexus. Ecological Engineering. 2015;**82**:330-335. DOI: 10.1016/j.ecoleng.2015.05.027

[12] Pandey VC. Assisted Phytoremediation. Amsterdam: Elsevier; 2021. p. 419

[13] Pandey VC, Gajic G. Green technologies for soil remediation. 2022;**108**:387-288. DOI: 10.1007/s00128-022-03485-8

[14] Chan-Quijano JG, Cach-Pérez MJ, López-Jiménez LN. Retos a futuro en el manejo sostenible de los recursos naturales. In: Chan-Quijano JG, Cach-Pérez MJ, editors. Manejo Sostenible de los Recursos Naturales: Experiencias y Retos a Futuro. Jalisco, México: Editorial Folia; 2023. pp. 315-354

[15] García-Chiang A. Acciones de responsabilidad social territorializadas en la industria del petróleo. Las cláusulas sociales en los nuevos contratos petroleros: ¿alternativa real para financiar el desarrollo local? In: Checa-Artasu MM, Hernández-Franyuti R, editors. El petróleo en México y sus impactos sobre el territorio. México: Instituto de Investigaciones Dr. José María Luis Mora; 2016. pp. 139-175

[16] Riojas C. Relaciones regionales: Uso y gestión de recursos naturales. In: Eufracio-Jaramillo JE, editor. Territorios en Conflicto: Ensayos Acerca de Disputas Socioambientales. Jalisco, México: El Colegio de Jalisco; 2016. pp. 29-50

[17] Ceccon E. La importancia del capital social en proyectos participativos de restauración ecológica. Letras Verdes. 2024:35(marzo-agosto). DOI: 10.17141/letrasverdes.35.2024.6058

[18] FAO, y UNEP. Global assessment of soil pollution: Summary for policy makers. 1st ed. Rome: Food and agriculture Organization of the United Nations. United Nations Environment Programme. 2021. 57 p. DOI: 10.4060/cb4827en

[19] Wackernagel M, Hanscom L, Jayasinghe P, Lin D, Murthy A, Neill E, et al. The importance of resource security for poverty eradication. Nature Sustainability. 2021;**4**:731-738. DOI: 10.1038/s41893-021-00708-4

[20] Toledo VM. Ecología, Sustentabilidad y manejo de recursos naturales: La investigación científica a debate. In: Oyama K, Manejo CA, editors. Conservación y Restauración de Recursos Naturales en México. México, Siglo XXI Editores: Universidad Nacional Autónoma de México; 2006. pp. 27-42

[21] Brown L. World on the Edge: How to Prevent Environmental and Economic Collapse. London: Routledge; 2011. 144 p. DOI: 10.4324/9781849775205

[22] Farinós-Dasí J. Gobernanza territorial para el desarrollo sostenible: Estado de la cuestión y agenda. Boletín de la Asociación Española de Geografía. 2008;**46**:11-32

[23] Brenner L, Rosales-Ortega R. Introducción. Procesos de construcción de gobernanza. In: Rosales-Ortega R, Brenner L, editors. Geografía de la Gobernanza: Dinámicas Multiescalares de los Procesos Económico-ambientales. México: Universidad Autónoma Metropolitana, Siglo XXI Editores; 2015. pp. 7-19

[24] Farinós-Dasí J. Desarrollo territorial y gobernanza: Refinado significados desde el debate teórico pensando en la práctica. Un intento de aproximación fronética. Desenvolvimiento Regional em debate. 2015;**5**:4-24

[25] Gjaltema J, Biesbroek R, Termeer K. From government to governance… to meta-governance: A systematic literature review. Public Management Review. 2020;**22**:1760-1780. DOI: 10.1080/14719037.2019.1648697

[26] Chan-Quijano JG, Domené-Painenao OE, Cach-Pérez MJ, Pat-Canche MK. Gobernanza ambiental y manejo de recursos naturales en la construcción de políticas públicas participativas para el desarrollo sostenible. In: Chan-Quijano JG, Cach-Pérez MJ, editors. Manejo

Sostenible de los Recursos Naturales: Experiencias y Retos a Futuro. Jalisco, México: Editorial Folia; 2023. pp. 239-268

[27] Zurbriggen C. Gobernanza: Una mirada desde América Latina. Perfiles Latinoamericanos. 2011;**19**:39-64

[28] Malena C. Good governance and civil society. In: List R, Anheier H, Toepler S, editors. International Encyclopedia of Civil Society. USA: Springer, Cham; 2020. pp. 1-9. DOI: 10.1007/978-3-319-99675-2_554-1

[29] Ken-Rodríguez CA. El desarrollo sostenible y regional en Othón P. Blanco, Quintana Roo y sus perspectivas para el desarrollo local. In: Chan-Quijano JG, Martínez-Yáñez R, Espinoza-Tenorio A, editors. Manejo Sostenible de los Recursos Naturales: Conservación y Experiencias. Chiapas, México: El Colegio de la Frontera Sur; 2012. pp. 101-123

[30] Huerta-Silva MH, Parra-Vázquez MR, Ianni E. Diseño de una política territorial de conservación y desarrollo basada en la sustentabilidad. El caso del municipio de Villaflores, Chiapas, México. Sociedad y Ambiente. 2024;**27**:1-34. DOI: 10.31840/sya.vi.2780

[31] Bejarano-González F. La Espiral del Veneno: Guía Crítica Ciudadana Sobre Plaguicidas. México: Red de Acción sobre Plaguicidas y Alternativas en México; 2007. 206 p

[32] Head BW. Community engagement: Participation on whose terms? Australian Journal of Political Science. 2007;**42**:441-454. DOI: 10.1080/10361140701513570

[33] Chirenje LI, Giliba RA, Musamba EB. Local communities' participation in decision-making processes through planning and budgeting in African countries. Chinese Journal of Population Resources and Environment. 2013;**11**:10-16. DOI: 10.1080/10042857.2013.777198

[34] Liu YR, van der Heijden MG, Riedo J, Sanz-Lazaro C, Eldridge DJ, Bastida F, et al. Soil contamination in nearby natural areas mirrors that in urban greenspaces worldwide. Nature Communication. 2023;**14**:1706. DOI: 10.1038/s41467-023-37428-6

[35] Seefoó-Luján JL. La Calidad es Nuestra, La Intoxicación... ¡de usted! Michoacán, México: El Colegio de Michoacán; 2005. p. 343

[36] Alarcón A, Ferrera-Cerrato R. Biorremediación de Suelos y Aguas Contaminadas Con Compuestos Orgánicos e Inorgánicos. México: Editorial Trillas; 2013. p. 333

[37] Jiménez-Cisneros BE. La contaminación Ambiental en México: Causas, Efectos y Tecnología Apropiada. México: Editorial Limusa; 2008. 928 p

[38] Tyagi B, Kumar N. Bioremediation: Principles and applications in environmental management. In: Saxena G, Kumar V, Shah MP, editors. Bioremediation for Environmental Sustainability: Toxicity, Mechanisms of Contaminants Degradation, Detoxification, and Challenges. Amsterdam: Elsevier; 2021. pp. 3-28. DOI: 10.1016/B978-0-12-820524-2.00001-8

[39] Aledo-Tur A, Domínguez-Gómez JA. Sociología Ambiental. España: Grupo Editorial Universitario; 2001. 439 p

[40] Ceccon E. Restauración en Bosques Tropicales: Fundamentos Ecológicos, Prácticos y Sociales. México: Universidad Nacional Autónoma de México, Ediciones Díaz de Santos; 2013. 283 p

[41] Chan-Quijano JG, Jarquín-Sánchez A, Ochoa-Gaona S,

Martínez-Zurimendi P, López-Jiménez LN, Lázaro-Vázquez A. Directrices para la remediación de suelos contaminados con hidrocarburos. Teoría y Praxis. 2015;**17**:123-144. DOI: 10.22403/UQROOMX/TYP17/05

[42] Provencio-Durazo E. Políticas públicas y medio ambiente. [Internet]. 2024. Available from: https://www.eventossccj.scjn.gob.mx/course/view.php?id=1567 [Accessed: May 13, 2024]

[43] Zhang S, Han Y, Peng J, Chen Y, Zhan L, Li J. Human health risk assessment for contaminated sites: A retrospective review. Environment International. 2023;**171**:107700. DOI: 10.1016/j.envint.2022.107700

[44] De Sousa-Santos B. Una Epistemología Del Sur. Siglo XXI Editores, Consejo Latinoamericano de Ciencias Sociales: México; 2009. 363 p

[45] Castro-Salazar JI, Camacho-García MO. Ideologías Ambientales en el Manejo y Conservación de la Naturaleza: Un Análisis Del Proceso Parlamentario y la Implementación de la Legislación Federal Mexicana. Ciudad de México: Tirant Lo Blanch; 2021. 253 p

[46] Leff E. Racionalidad Ambiental: La Reapropiación Social de la Naturaleza. México: Siglo XXI Editores; 2004. 479 p

[47] Shiva V. Manifiesto Para Una Democracia de la Tierra: Justicia, Sostenibilidad y Paz. Barcelona, España: Paidós Ibérica; 2006. 229 p

[48] Lee H, Sam K, Coulon F, De Gisi S, Notarnicola M, Labianca C. Recent developments and prospects of sustainable remediation treatments for major contaminants in soil: A review. Science of the Total Environment. 2024;**912**:168769. DOI: 10.1016/j.scitotenv.2023.168769

[49] Chan-Quijano JG, Caamaño-Urgell PC, Vázquez-Asencio M, Gutiérrez-Torres H, Winzig-Gómez TI. Sostenibilidad y políticas públicas ante la problemática socioambiental de los derrames de petróleo: Caso de estudio en huertos familiares. In: Chan-Quijano JG, Cach-Pérez MJ, editors. Manejo Sostenible de los Recursos Naturales: Experiencias y Retos a Futuro. Jalisco, México: Editorial Folia; 2023. pp. 269-313

[50] Hou D, Al-Tabbaa A, O'Connor D, Hu Q, Zhu YG, Wang L, et al. Sustainable remediation, and redevelopment of brownfield sites. Nature Reviews Earth & Environment. 2023;**4**:271-286. DOI: 10.1038/s43017-023-00404-1

[51] Nurse A. Green criminology: Shining a critical lens on environmental harm. Palgrave Communications. 2017;**3**:10. DOI: 10.1057/s41599-017-0007-2

[52] Lynch MJ. Green criminology and environmental crime: Criminology that matters in the age of global ecological collapse. Journal of White Collar and Corporate Crime. 2020;**1**:50-61. DOI: 10.1177/2631309X19876930

[53] Carpio-Domínguez JL, Criminología verde para México. Desarrollo de Una Perspectiva Verde en la Herencia Criminológica Mexicana. México: Universidad Autónoma de Tamaulipas, Tirant lo Blanch; 2023. p. 255

[54] Leff E. El Conflicto de la vida. Ciudad de México: Siglo XXI Editores; 2020. p. 394

[55] Hannam I, Boer B. Legal and Institutional Frameworks for Sustainable Soils: A Preliminary Report. Gland, Switzerland and Cambridge, UK: International Union for Conservation of Nature Resources; 2002. p. 88

[56] Azam MM. Soil contamination and remediation measures: Revisiting

the relevant laws and institutions. In: Hasegawa H, Rahman I, Rahman M, editors. Environmental Remediation Technologies for Metal-Contaminated Soils. Tokyo: Springer; 2016. pp. 99-124. DOI: 10.1007/978-4-431-55759-3_5

[57] Meza-García AC, Chan-Quijano JG. Ecological risk index, implications, and its bioconcentration ratio in soils contaminated with heavy metals. In: Mustafa A, Naveed M, editors. Soil Contamination: Recent Advances and Future Perspectives. London, United Kingdom: IntechOpen; 2022. pp. 93-104. DOI: 10.5772/intechopen.108801

[58] Hou D, O'Connor D, Jin Y. Sustainable remediation, and socio-environmental management at contaminated sites. In: Ok YS, Rinklebe J, Hou D, Tsang DCW, Tack FMG, editors. Soil and Groundwater Remediation Technologies: A Practical Guide. Boca Raton: CRC Press; 2020. p. 9. DOI: 10.1201/9780429322563

[59] Grifoni M, Franchi E, Fusini D, Vocciante M, Barbafieri M, Pedron F, et al. Soil remediation: Towards a resilient and adaptive approach to deal with the ever-changing environmental challenges. Environments. 2022;**9**(2):18. DOI: 10.3390/environments9020018

[60] Diver S, Blaich Vaughan M, Baker-Medard M. Collaborative care in environmental governance: Restoring reciprocal relations and community self-determination. Ecology and Society. 2023;**29**:7. DOI: 10.5751/ES-14488-290107

[61] Cohen-Shacham E, Walters G, Janzen C, Maginnis S. Nature-Based Solutions to Address Global Societal Challenges. Gland, Switzerland: International Union for Conservation of Nature Resources; 2016. p. 97

[62] Challenger A, Córdova A, Lazos-Chavero E, Equihua M, Maas M. Opportunities and obstacles to socioecosystem-based environmental policy in Mexico: Expert opinion at the science-policy interface. Ecology and Society. 2018;**23**:31. DOI: 10.5751/ES-10066-230231

[63] Lazos-Chavero E. Introducción: Experiencias que enriquecen las conceptualizaciones y las luchas en la defensa de los comunes en América Latina. In: Lazos-Chavero E, editor. Retos Latinoamericanos en la Lucha por los Comunes: Historias a Compartir. Buenos Aires, Argentina: Consejo Latinoamericano de Ciencias Sociales; 2020. pp. 11-36

[64] Jonsson A, Haller H. Sustainability aspects of in-aitu bioremediation of polluted soil in developing countries and remote regions. In: Hernández-Soriano MC, editor. Environmental Risk Assessment of Soil Contamination. London, United Kingdom: IntechOpen; 2014. pp. 57-86. DOI: 10.5772/57315

[65] SERMANAT (Secretaría de Medio Ambiente y Recursos Naturales). Programa Nacional de Remediación de Sitios Contaminados. México: Secretaría de Medio Ambiente y Recursos Naturales; 2019-2024. p. 44

[66] Yasutaka T, Zhang H, Murayama K, Hama Y, Tsukada Y, Furukawa Y. Development of a green remediation tool in Japan. Science of the Total Environment. 2016;**563-564**:813-821. DOI: 10.1016/j.scitotenv.2016.01.018

[67] MOE (Ministry of the Environment Government of Japan). IAEA Assistance to the Ministry of the Environment, Japan on 'Volume Reduction and Recycling of Removed Soil Arising from Decontamination Activities after the Accident of the Fukushima Daiichi Nuclear Power Station'. Japan: Ministry of

the Environment Government of Japan; 2023. p. 30

[68] EPA (United States Environmental Protection Agency). Green Remediation Best Management Practices: Mining Sites EPA 542-F-12-028. United States of America: United States Environmental Protection Agency; 2012. p. 6

Chapter 5

Managing Drill Cuttings Waste in Our Age

Mary Allagoa

Abstract

Oil and gas drilling processes produce drill cutting wastes around the world. Drill cuttings are formed from the drill well bore to the earth's surface and differ from a fine to pebble size rock. They carry with them petroleum hydrocarbons; water, and various drill mud. Drill cuttings have been treated with several methods over the years. The method include, thermal technology, solidification/stabilization, bio remediation and mechanochemical method. This chapter explains few techniques employed in drill cuttings treated waste, which may be convenient and affordable. This will heighten the need for bio-treating drill cuttings before disposal to ease the level of environmental pollution.

Keywords: management, sustainability, pollution, socio-environmental, public policies

1. Introduction

Oil and gas operations create drill cutting wastes around the world [1]. Before now, these drill cuttings were usually dumped into water bodies or land without recourse to any form of treatment [2]. Drill cuttings refer to any material (solids) removed from a borehole while drilling petroleum wells. Drill cuttings are granular in nature, and constitute solid phase of the waste stream [3]. The chemical characteristics of the drill cuttings depend to a greater extent on the chemistry of the drilling fluid and mud additives used in the drilling process [4]. Drilling muds are liquids used in the drilling process, which comprises of water, oil or synthetic based. Drilling muds are essentially made of oil or water, ground rock and clays [5]. Studies has shown that synthetic muds are not breaking down naturally in sea water as quickly as expected and are most harmful because they contain large quantities of hydrocarbon. Drill cuttings are characterised by relatively high content of polycyclic aromatic hydrocarbon and heavy metals [6]. PAH has negative influence on animal and plant organisms. PAHs have been discussed in numerous research works as environmentally harmful because they are or can become carcinogenic or mutagenic [7–10]. According to the international center for soils and contaminated sites [11]. PAHs are defined by their high durability in the environs, which allows them to store in the soil for numerous period and degrade with difficulty. Because of this, there is need to have these drill cuttings treated before disposal.

IntechOpen

1.1 Treatment method of drill cutting

1.1.1 Thermal treatment technologies

Thermal technologies use high temperatures to reclaim or destroy contaminated material. They can be classified into two class (**Figure 1**) [12]. The first function by the use of incineration to destruct hydrocarbons by heating them to very higher temperatures (1200–1500°C) in the existence of air. Incineration is not normally used for drilling waste material but has greater relevance for materials like medical waste material and Convert them into less bulky materials that are non- hazardous or less hazardous prior to incineration [13]. The second group called thermal desorption uses a non- oxidizing process to vaporize volatile and semi-volatile through the application of heat. Of the various technologies involved in thermal desorption, thermal phase separation (TPS) is the most popular and consist of five subsystems [14]. The five subsystems are pre-treatment system; the anaerobic thermal desorption unit (ATDU), solid discharge and conditioned sub system, vapor recovery unit (VRU), and finally water treatment unit (WTU). Thermal desorption technologies rely on volatilization; treatment ratio is connected to the volatility of the toxins. Hence thermal desorption well removes Low Molecular Weight (LMW) hydrocarbons, aromatic and other volatile organics while High Molecular Weight (HMW) PAHs are less easily separated. Costs of thermal treatments are quite prohibitive [15], personnel and equipment is exposed to the resulting fugitive dusts.

1.1.2 Solidification/soldification

These involve the mechanical binding of the cuttings to form a solid block as a result of the chemical interaction between the cuttings, the solidifying and solidifying agents [7–10]. These approaches are aimed at ensuring that residual oil and heavy

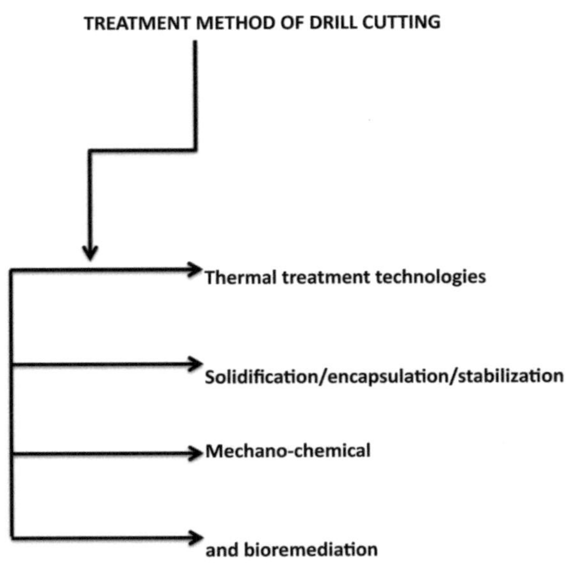

Figure 1.
Drill cutting treatment methods.

metals in the pre- treated, mud coated cuttings are prevented from leaching to allow for their land application or their use as road construction materials. Al-Ansary and Al-Tabbaa [16] studied Stabilization/solidification of synthetic petroleum drill cuttings. They ascertained the leachability results and display the reduction of the synthetic drill cuttings to a stable non-reactive hazardous waste, defiant with the UK approval criteria for non-hazardous landfills. Boutammine et al. [17] uses stabilization/solidification (S/S) and combined with biological treatment on drill cuttings. The experimental solidification tests were analyzed. The results display at 28 days of curing show the UCS was importantly affected by additives. A similar exercise carried out in Southern Louisiana University indicated that cuttings stabilized in a silica matrix had a pH > 11 and did not support plant growth when land applied [18].

1.1.3 Mechanochemical (MC)

Mechano-chemical is a method projected since 1902. Presently is effectively exploited in different fields. It has also been recommended in the past for cleaning organic wastes material. Mechanochemistry explains the mechanical breakage of intramolecular bonds by external force, where contacts between micronized molecular solids are defined by the mechanical action for common motion. Mechanochemical treatment allows chemically stable galactic organic unit to act at comparatively moderate operating conditions owing to mechanical energy transferred when grinding [19]. Processes includes; Grinding, milling, shearing, scratching, polishing, and rapid friction which provide the mechanical impact for mechanochemistry. The following are factors affecting the milling process [19]. These are; Soil composition and characteristics, Properties of the contaminant, Optimum milling duration and Ball to soil ratio. Mechanochemistry covers solid-state reactions by friction at lubrication of rapidly moving cold contacting surfaces, and single bond breaking or cutting [20] (See below) for brief description (**Figure 2**).

1.1.4 Bioremediation technology

It could be called bio-restoration, meaning giving nature a helping hand. Bioremediation can be characterized as any procedure that uses micro- organisms to decontaminants pollutant. To its original condition [21–24]. Example are land farming, composting, bioreactors, bioventing, bio filters, bio-augmentation, bio stimulation, remediation by enhanced natural attenuation (RENA) [25].

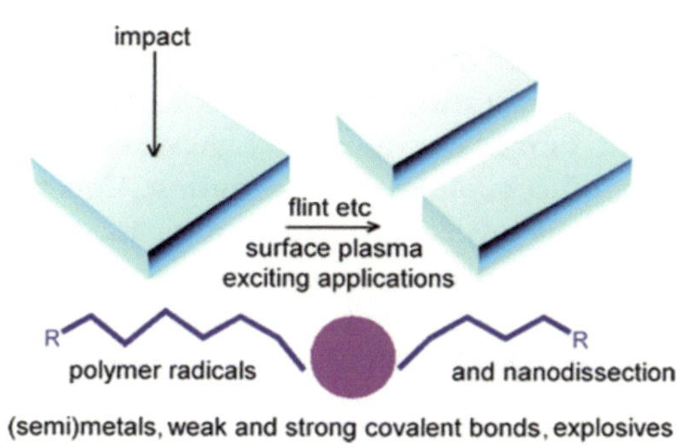

Figure 2.
Processes involved in this treatment [20].

- Bio stimulation- is the process of introducing additional nutrients in the form of organic and/or inorganic fertilizers into contaminated system. These requirements can gradually be satisfied by addition of nitrate, phosphate and sulphate containing salts.

- Bio venting- method of treating contaminated soils by drawing oxygen through the soil to stimulate microbial activities.

- Bio reactors: container or reactor may be used to treat liquids or slurries.

- Bio augmentation: involves bacterial cultures to a contaminated medium. It also involves the introduction of microorganisms (cultured) into a contaminated system. These practices take care of the events of the absence of competent naturally occurring microbes. Although some of these specialize microbes are naturally occurring; others can be synthesize in the laboratory [21–24].

- Remediation by enhanced natural attenuation (RENA) - is an in situ strategy that involves the combined application of bio stimulation with inorganic fertilization and certain agro technical processes like tilling in windrow to decontaminate a polluted matrix. Bio-stimulation helps to increase the activities of indigenous heterotrophic bacteria (HTB) and hydrocarbon – utilizing bacteria (HUB) while tilling helps to improve aeration [21–24].

- Composting is a involves aerobic conditions in a matric phase environment. It is a successful strategy for the successful recycling of organic wastes [21–24]. Composting is an aerobic process consisting of aerating sludge mixed with co-compost such as saw dust and animal manure. Composting is increasing being the preferred methods of treatment of municipal sludge's because the process produces marketable end products that can be used as a container and organic fertilizer.

Composting involves the addition of organic matter to promote the development of a wide range of microbes to breakdown complex contaminant under a good temperature, moisture and nutrient level [21–24].

Once after initiating composting its proceeds in two phase namely; high rate phase and curing phase. During high rate phase, thermophilic temperature (50°C and above) are reached owning to increased microbial activity and heat generation by biodegradable component in the compost. The curing phase is a low temperature stage, which follows after the rapidly degradable components are utilized. After these stages, the material will stabilized and is ready for land application. Meanwhile, to maintain the thermophilic temperature range, periodic turning is necessary [21–25]. Composting is divided into four major micro biological stages which are; temperature, mesophilic, thermophilic, cooling and maturation. Composting had some viable advantages including relatively low capital and maintenance cost, simple design and operation, and some removal of oil pollution. Composting consist of three major types of feed; the material been degraded (industrial waste), amendment (easily degraded organics; nutrients and microbes) and bulking agents (such as wood chips, saws dust, rice hulls, farmyard manure used for moisture control). Bulking agent are materials of low density when with soils; reduce the soil bulk density, produce higher porosity, may produce higher oxygen which may form water stable aggregates [21–25]. These alteration to a soil will gain aeration and microbial action. Too much moisture may impede thermophilic condition and too little moisture can seriously reduce reaction rate. As a result of these processes, nitro aromatic are converted into nontoxic end product such as mono and di – amino toluene and carbon-dioxide.

- Phytoremediation: uses plant and micro-organisms to make contaminant harmless [26]. Although, report by [27] states that phytoremediation involves the translocation and transpiration.

2. Materials and methods

2.1 Laboratory analysis

The samples was air-dried and weighed. The ratio of samples to the solvent was 1:2. The required quantity of analytical grade hexane and dichloromethane. The dry/cold extraction was applied and gas chromatography machine used (**Table 1**).

2.2 Determine PAHs sources

PAH ratios determine PAH sources, clarify samples by locations, and estimate (**Table 2**) [30].

2.3 Benzo[a]Pyrene equivalent (B[a]Peq) estimation

BaP equivalent concentration (BaPeq) evaluated the toxicities of PAHs in sampling sites. Therefore, the total PAH concentration is expressed as B[a]Peq to illustrate the toxic potency [31]. As proposed earlier by [31, 32], the B[a]Peq is the summation of the B[a]Peqi. It is the value for specific PAHs or individual PAH concentrations in the sample (cPAHi) multiplied by its toxic equivalency factor (TEFPAHi).

$$B[a]Peq = \Sigma(BaPeqi) = \Sigma(cPAHi \times TEFPAHi) \ or \ \Sigma BaPeq = \Sigma Ci \times TEFi \quad (1)$$

Where Ci is the concentration of individual PAHs, TEFi is the corresponding toxic equivalency factor (**Table 3**).

Reactor	Compost
Control	Untreated drill-cuttings
SAMPLE A	Drill cuttings + top soil + (PD + SD) i.e. 4:1:1; 2000 g + 500 g + 500 g
SAMPLE B	Drill cuttings + top soil + SMS i.e. 4:1:1; 2000 g + 500 g + 500 g
SAMPLE C	Drill cuttings + top soil + (PD + SD) i.e. 4:1:2; 2000 g + 500 g + 1000 g
SAMPLE D	Drill cuttings + top soil + SMS i.e. 4:1:2; 2000 g + 500 g + 1000 g
SAMPLE E	Drill cuttings + top soil + (PD + SD) i.e. 4:1:4; 2000 g + 500 g + 2000 g
SAMPLE F	Drill cuttings + top soil + SMS i.e. 4:1:4; 2000 g + 500 g + 2000 g
SAMPLE G	Drill cuttings + top soil + (PD + SD) + SMS i.e. 4:1:1; 2000 g + 500 g + 500 g

PD – Poultry droppings, SD – Saw dust, SMS – Spent mushroom substrate.

Table 1.
Experimental layout.

PAHs ratio	Values	Source	Reference
∑LMW/∑HMW	<1 >1	Pyrogenic/Anthropogenic Petrogenic/natural	[28]
Ant/(ant + Phe)	<0.1 >0.1	Petrogenic/natural Pyrogenic/Anthropogenic	[29]
BaA/(BaA + CHR)	<0.2 0.35	Petrogenic/natural Combustion	[30]

Table 2.
Diagnostic ratios used in this study with their typical values for particular processes.

PAHs	Toxicity equivalent factor	Reference
Naphthalene	0.001	[32]
Phenanthrene	0.001	
Anthracene	0.01	
Acenaphthelene	0.001	
Acenaphthylene	0.001	
Flourene	0.001	
Pyrene	0.001	
Chrysene	0.01	
Benzo[a]anthracene	0.1	
Fluoranthane	0.001	

Table 3.
Toxicity equivalent factor value of the individual PAHs.

2.4 Quantification and characterization of degradation

The biodegradation rates of PAHs were evaluated by comparing the reaction rate constants of the pseudo-first-order kinetics refereed to [9].

$$\log Co - Ct = \log Co - \left(\frac{K1}{2.303}\right) * t \tag{2}$$

Make K_1 the subject formula

$$K1 = \frac{2.303^*}{t}\left[(\log Co) - (\log Co - Ct)\right] \tag{3}$$

K - The apparent constant reaction rate of the pseudo-first-order (1/week), t-Time (weeks).

Then, the half-life of the respective PAHs:

$T1/2 = 0.693/K1$ (3).

Biodegradation efficiency **(BDE)**:

$$BDE(\%) = \frac{Co - Ct^{*}}{Co} 100 \tag{4}$$

Where C_o – initial concentration C_t – final concentration.

2.5 Statistical analysis

The data were presented as the mean of triplicates (n = 3) ± standard error. ANOVA or general linear model (GLM) tests in MINITAB 16.0 was identified as $p \leq 0.05$.

3. Results

The ∑14PAHs reduced at a reasonable rate. Sample A at 4 weeks had ∑14PAHs to be 12747.4 mg/kg which reduced to 9107.9 mg/kg at about 28.6%. Sample B with 11195.5 mg/kg reduced to 3912.0 mg/kg at 65.1% while Sample C had 11646.5 mg/kg which reduced to 6296.8 mg/kg at about 45.9%. And Sample D reduced from 10897.2 to 4704.5 mg/kg at 56.8% while Sample E had a 50.2% with a value of 6113.0 to 3041.7 mg/kg after 4 and 8 weeks. Sample F & G reduced at 53.6 & 48.3% respectively. The concentration reduced from 4266.4 to 1978.7 mg/kg and 7543.2 to 3898.6 mg/kg in sample F & G respectively. For Bio degradation Efficiency (BDE), the % ranges from 11.1–86.0% in Sample A, Sample B had 15.6–90.6% while Sample C had 18.5–95.8%. Yet, 27.5–81.9% were observed for sample D and 27.9–90.7% in sample E. Additionally, Sample F (25.4–82.2%) while sample G (8.5–89.6%) respectively. For the individual PAHs, B[a]Ant had the lowest BDE in Sample A & D while PHE had the lowest BDE in Sample C and F. The PAHs Indeno [1–3] ANT, Indeno [1,2,3-cd]PY and FLUROANT were the lowest BDE in sample B, E and G respectively. The highest BDE values were observed for B[b]FLOURANT in Sample C and E while Sample A,B,D,F & G show CHRY, B[A]PY, FLUORNAT, DIBENZ[A,H] ANT & INDENO[1,2,3]ANT with the highest BDE respectively. The PAHs in this study were group on ring bases. Where NAP (∑2 ring), ACE, ACY, FLUORENE, PHE & ANT as ∑3 rings. FLUORANT, B[A]ANT & CHRYS were ∑4 rings and B[b] fluornt, B[A]PY, INDENO(1,2,3)ANT, DIBENZ(a,h)ANT as ∑5 rings. In Addition, INDENO [1,2,3-cd] PYR is ∑6 ring. The rings were 2,3,4, 5 and 6 rings respectively. As seen, at 4 weeks ring ∑2 (0.16%), ∑3 ring (29.60%), ∑4 ring (30.63%), ∑5 ring (16.62%) and ∑6 ring (22.99%). Further at 8 weeks we have % of the rings to be ∑2 rings (0.16%), ∑3 rings (32.19%), ∑ 4 rings (34.10%), ∑5 rings (11.10%) and ∑6 rings (22.44%). Pearson correlation at 4 and 8 weeks show a value of 0.847 and there was a highly significant different with the amendment at 8 weeks as $p < 0.05$ (**Figure 3**; **Tables 4–6**).

To further investigate possible sources of PAH pollution, Principal Component Analysis was performed using the correlation matrix of the log-transformed PAH levels. One principal components PC1 with eigenvalue >1 were extracted and explained 88.8% (High loading variables are presented). PC1 was characterized by high loadings of PAHs with 2–3 ring, 4 ring, and 5–6 ring. The 14 PAHs had high loading where the variable depict the seven amendments. Thus, PC1 had a proportion of 88.8% of the

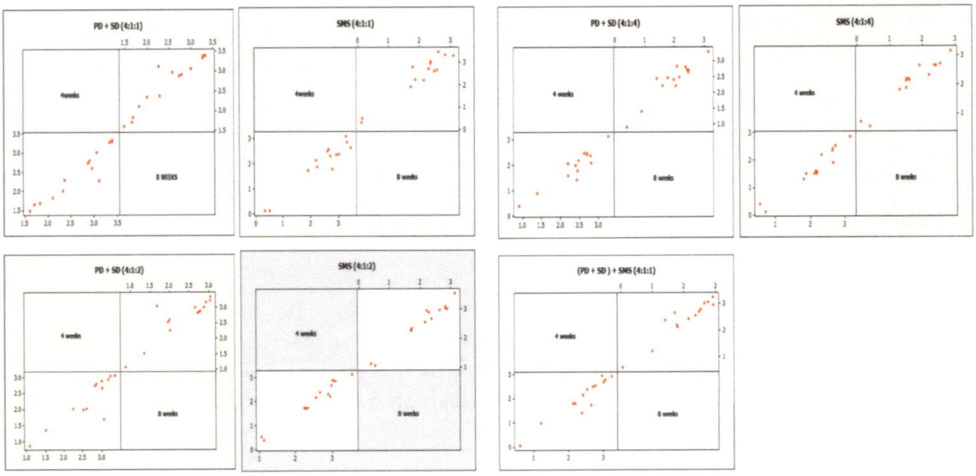

Figure 3.
Show an evenly distribution of the PAHs values after normality test at 4 and 8 weeks transformed with log 10.

4 Weeks (%)					
COMPOST	2	3	4	5	6
1	0.39	48.82	38.43	5.11	7.25
2	0.02	33.39	28.69	12.28	25.62
3	0.27	30.02	40.64	20.41	8.66
4	0.10	25.24	27.36	15.07	32.22
5	0.13	21.17	25.86	22.00	30.83
6	0.12	21.12	19.67	25.93	33.17
8 Weeks (%)					
1	0.48	56.65	34.21	4.42	4.23
2	0.03	34.01	44.40	11.17	10.39
3	0.35	37.21	38.19	12.19	12.06
4	0.07	27.34	33.44	11.05	28.09
5	0.08	18.04	25.72	11.49	44.67
6	0.07	24.26	21.00	18.04	36.64

Table 4.
Rings % distribution on and after bio-remediation.

total variance and reflected the contribution of anthropogenic to the origin of PAHs at 4 weeks. At 8 weeks, Eigenvalue was 5.8522 greater than 1 hence only PC1 was described by high loading of PAHs with a proportion of 83.6% of the total variance and reflected the contribution of anthropogenic activities to the origin of PAHs. The PCA results, in combination with diagnostic ratios, suggested that anthropogenic sources were probably the main sources. PAH ratios were calculated to determine PAH sources. The calculated ratios were ∑LMW/∑HMW, Ant/ (Ant+ Phe), and BaA/ (BaA + Chry). Their values were < 1, >0.1, and > 0.35 respectively. Sample A toxicity values reduced

Variable	PC1	PC2	PC3	PC4	PC5	PC6	PC7
1	0.336	−0.821	0.329	−0.134	0.057	0.090	0.273
2	0.392	0.096	0.499	−0.045	−0.117	−0.318	−0.687
3	0.380	−0.202	−0.468	0.709	−0.100	−0.286	−0.048
4	0.392	0.040	−0.295	−0.091	0.435	0.658	−0.357
5	0.385	0.270	−0.087	−0.288	0.557	−0.503	0.351
6	0.374	0.441	0.437	0.359	−0.179	0.350	0.437
7	0.383	0.078	−0.373	−0.508	−0.664	0.021	0.095
Eigenvalue	**6.2178**	0.4170	0.1308	0.1072	0.0882	0.0277	0.0114
Proportion	**0.888**	0.060	0.019	0.015	0.013	0.004	0.002
Cumulative	**0.888**	0.948	0.967	0.982	0.994	0.998	1.000

Table 5.
Eigen-analysis of the correlation matrix (4 weeks).

Variable	PC1	PC2	PC3	PC4	PC5	PC6	PC
1	0.330	−0.814	−0.242	−0.129	0.058	0.340	0.184
2	0.396	0.010	−0.121	0.319	−0.736	−0.341	0.261
3	0.387	−0.047	−0.034	0.698	0.385	−0.067	−0.455
4	0.387	0.065	0.632	−0.028	0.352	−0.138	0.550
5	0.383	0.434	0.070	−0.041	−0.206	0.781	−0.077
6	0.369	0.360	−0.661	−0.316	0.341	−0.232	0.164
7	0.390	−0.108	0.290	−0.541	−0.153	−0.284	−0.596
Eigenvalue	**5.8522**	0.5088	0.2354	0.2092	0.0835	0.0639	0.0470
Proportion	**0.836**	0.073	0.034	0.030	0.012	0.009	0.007
Cumulative	**0.836**	0.909	0.942	0.972	0.984	0.993	1.000

Table 6.
Eigen-analysis of the correlation matrix (8 weeks).

from 281.8 to 191.5 of 32.0% while Sample B, C, D, E, F & G toxicity potency values were 86.0, 69.2, 56.9, 57.8, 50.3 and 82.2%. In Addition, Sample B, C, D, E, F and G values were reduced from 679.4 to 94.5 Ba mgPeq kg, 543.0 to 197.2, 316.9 to 136.6, 304.1 to 128.2, 467.0 to 232.2 and 532.4 to 94.9 respectively. The samples had the following values of k and $t_{1/2}$: Sample A (k 0.005–0.655 week $^{-1}$ & $T_{1/2}$ 1.1–148.7 week), Sample B (k 0.002–0.278 week $^{-1}$ & $T_{1/2}$ 2.5–283.9 week), Sample C (k 0.004–0.446 week $^{-1}$ & $T_{1/2}$ 1.4–194.4 week), Sample D (k 0.003–0.148 week $^{-1}$ & $T_{1/2}$ 4.7–211.1 week), Sample E (k 0.009–0.155 week $^{-1}$ & $T_{1/2}$ 4.5–78.5 week), Sample F (k 0.010–0.807 week $^{-1}$ & $T_{1/2}$ 0.9–72.6 week) and Sample G (k 0.007–0.343 week $^{-1}$ & $T_{1/2}$ 2.0–96.0 week).

4. Discussion

PAHs are of great concern due to their documented carcinogenicity and endocrine disruptive activity [7–10]. The 5 and 6-ring PAHs had degraded the least in this study, these may be due to sequestration in the compost matrix [33]. Co-composting of PAH-contaminated soil with poultry manure was investigated by [34]. Studies by Refs. [35–38]; established that high molecular weight PAHs are more recalcitrant in the environment and may resist both chemical and microbial degradation. Nevertheless, Composting not only promotes the growth of plants but also enhances the growth and activities of soil microbes [38]. Poultry droppings (PD), when added may therefore serve as a nutrient supplement to the soil microbial population needed to stimulate biodegradation of the total and poly-aromatic hydrocarbons, thereby ameliorating the risk these compounds pose to environment [39]. Some studies [40–42]; found greater disappearance rates after inoculating their samples with white rot fungi. In this study, SMS was able to degrade a significant amount of 3, 4, 5, and 6 rings, which indicates the potential of SMS to degrade PAHs. A lot of substrates can encourage the fungal degradation of organic pollutants. Spent mushroom compost provides bulk nutrients for indigenous soil micro flora and contains considerable microbial activity [40–43]. These can be shown by [44], they tested the potential of a mixed substrate used for mushroom cultivation (MCS) in the bioremediation of aged, PAH-contaminated soil. Three PAHs, anthracene, benzo(a)pyrene and benzo(a)anthracene, were most subject to degradation, which is consistent with the PAH degradation features of fungal laccase. Since poultry manure is rich in carbon and mineral nutrient, particularly nitrogen [45] and SMS have the ability to degrade lignin and PAHs. A combination of poultry droppings and SMS degraded the PAHs better in the drill cuttings than when with one amendment.

Another study by [46], show a high degradation percentages of total hydrocarbons (82%), n-alkanes (96%) and the 16 USEPA-listed polycyclic aromatic hydrocarbons (93%), when applying 75 and 33% of organic wastes. Although, [46] suggest the application of 75% organic waste produced better result when compared to 33%. In general, the physicochemical properties are the main factor that determines the level of PAHs in the environment and soil sorption of the compound [47]. For the toxicity equivalent values in this study, the total BaPeq (Toxicity values) of 12PAHs in drill cuttings was higher than urban soil in Lisbon, Portugal (229 mg BaPeq/kg [48], Hunpu (52.31 μg/kg) [49], Xinzhou (34 μg/kg) [50], Liaohe estuary (30.0 μg/kg) and Yellow River Delta (11.92 μg/kg) [51] and lower in Palermo, Italy (151–4291 mg BaPeq/kg, [48]. Also, Canadian Soil Quality Guidelines for commonly occurring parent PAHs for the protection of environmental and human health provide PAH guidelines were based on the PAHs' carcinogenic effects [52]. Biodegradability is usually explained by first-order kinetics [39]. In Ref., [39] explained this by suggesting that the higher the biodegradation rate constants, the faster the rate of biodegradation, and consequently, the lower the half-life. The inconsistent effects of nutrient, positive or negative, were clarified by [53], who proposed a resource-ratio theory to envisage how competition for growth-limiting resources influenced biological diversity and function within a biotic community. Braddock et al. [54] recommended that the optimization of the biodegradation rate, by a specific nutrient ratio may differ as dissimilar PAH-degrading microorganisms require different ratios. The effects of biodegradation, the role, and ratio of different nutrients, and the selection of particular degraders, require further research [53].

5. Conclusion

Based on the results obtained in this study, the following conclusions are drawn;

- Using the two types of wastes (plant and animal) is more preferable. Since they both have different properties that have effect on the PAHs.

- Using more quantity of amendment is more effective to degrade PAHs as regard to this study.
- P.Ostreatus degrades PAHs better than poultry dropping. Since P. ostreatus is a non – invasive and non-pathogenic fungus, commonly grown and eaten. It may be more readily accepted by the public for bio-remedial application.
- For composting, large quantity of amendment may be better to enhanced degradation.

Drill cuttings values exceed the safe limit and therefore remain unsafe for land disposal without prior treatment.

6. Recommendation

This study therefore recommends that:

- Drill cuttings should be bio-treated, to bring the contaminant level to acceptable limits, before its land disposal to reduce the level of environmental pollution.
- Oil based mud should be banned while drilling by oil companies except the companies treat the drill cuttings before disposal.
- I recommend that composting should be used with two amendments in treating drill cuttings.
- The findings in this study should be adapted by exploration and production (E&P) waste management companies that are into drill cuttings treatment to reduce the enormous amounts of money, energy and pollution associated with thermal treatment technologies.

Acknowledgements

My acknowledgement goes to my siblings Dr. Theresa C. Allagoa (MD), Dr. Jennifer C. Allagoa (MD) and Josephine Allagoa (Registered Nurse) for their encouragement.

Dedication

I dedicate this chapter to my lovely mother Chief Hon. Mrs. Mary Allagoa. For her unending love, Care, Encouragement to me and my siblings.

Author details

Mary Allagoa
Environmental Soil Science, King's College, University of Aberdeen,
Aberdeen, Scotland, United Kingdom

*Address all correspondence to: maryallagoa@gmail.com

References

[1] Kazamias G, Zorpas AA. Drill cuttings waste management from oil & gas exploitation industries through end-of-waste criteria in the framework of circular economy strategy. Journal of Cleaner Production. 2021;**322**:129098

[2] Browning K, Seaton S. Drilling waste management: Case histories demonstrate that effective drilling waste management can reduce overall well-construction costs. In: SPE Annual Technical Conference and Exhibition, Dallas, Texas, October 2005. 2005. DOI: 10.2118/96775-MS. Paper Number: SPE-96775-MS, ISBN: 978-1-55563-150-5

[3] Njuguna J, Siddique S, Kwroffie LB, Piromrat S, Addae-Afoakwa K, Ekeh-Adegbotolu U, et al. The fate of waste drilling fluids from oil & gas industry activities in the exploration and production operations. Waste Management. 2022;**139**:362-380

[4] Environmental Protection Act (1994). Characterization and Management of Drilling Fluids and Cuttings in the Petroleum Industry. Environmental Protection Act; 1994. ESR/2016/2001, Version 2.01, Effective: 16 FEB 2024

[5] Ismail AR, Aliasc AH, Rosli W, Sulaimana W, Jaafara MZ, Ismaila I. Drilling fluid waste Management in Drilling for oil and gas Wells. The Italian Association of Chemical Engineering. 2017;**56**:2283-9216

[6] Allagoa M. Comparing the use of GGBS and MgO in laboratory and field trial mixing of contaminated made ground soil. In: GeoMontréal 2013: 66th Canadian Geotechnical Conference; 11th Joint with IAH-CNC. 2013

[7] Allagoa M. Effective use of cementitious materials, ground granulated blast furnace slag and bentonite in treating contaminants on polluted land. In: Proceedings of the Shale Energy Engineering 2014 Conference, Held in Pittsburgh, Pennsylvania, July 21-23, 2014. Sponsored by the Energy Division of ASCE; 2014

[8] Allagoa M. Feasibility of bioremediation technique on polycyclic aromatic hydrocarbon. In: World Environmental and Water Resources Congress 2024: Climate Change Impacts on the World we Live In. ASCE Press; 2023. pp. 362-372

[9] Allagoa M, Al-Tabbaa A. Dry Binder Mixing of Field Trial Investigation Using Soil Mix Technology: Case Study on Contaminated Site Soil. International Journal of Environmental and Ecological Engineering, World Academy of Science, Engineering and Technology. 2024;**18**(4):131-137

[10] International Centre for Soil and Contaminated Sites. Manual for Biological Remediation Techniques. International Centre for Soil and Contaminated Sites. Open Access and Academic Publisher; 2006

[11] Morillon A, Vidalie J-F, Hamzah US, Suripno S, Hadinoto EK. Drilling and waste management. In: Paper Presented at the SPE International Conference on Health, Safety and Environment in Oil and Gas Exploration and Production, Kuala Lumpur, Malaysia, March 2002. Paper Number: SPE-73931-MS; 2002. DOI: 10.2118/73931-MS

[12] Ding D, Song X, Wei C, LaChance J. A review on the sustainability of thermal treatment for contaminated soils. Environmental Pollution. 2019;**253**:449-463

[13] Zupan T, Kapila M. Thermal desorption of drill muds and cuttings in Ecuador: The environmental and financially sound solution. In: Paper Presented at the SPE International Conference on Health, Safety and Environment in Oil and Gas Exploration and Production, Stavanger, Norway, June 2000. Paper Number: SPE-61041-MS; 2000. DOI: 10.2118/61041-MS

[14] Bansal KM, Sugiarto. Exploration and production operations - waste management a comparative overview: US and Indonesia cases. In: Paper Presented at the SPE Asia Pacific Oil and Gas Conference and Exhibition, Jakarta, Indonesia, April 1999. Paper Number: SPE-54345-MS; 1999. DOI: 10.2118/54345-MS

[15] Shaffer GP, Hester MW, Miller S, Des Rochers DJ, Southern RF, Childers GW, Campo FM. Restored Drill Cuttings for Wetland Creation: Results of a Two-Year Mesocosm Approach to Emulate Field Conditions under Varying Hydrologic Regimes (No. DOE/BC/14849-3). Houston, Texas; 1998. Available from: https://digital.library.unt.edu/ark:/67531/metadc687884/m1/2/.

[16] Al-Ansary MS, Al-Tabbaa A. Stabilisation/solidification of synthetic petroleum drill cuttings. Journal of Hazardous Materials. 2007;**142**(2):410-421

[17] Boutammine H, Salem Z, Khodja M. Petroleum drill cuttings treatment using stabilization/solidification and biological process combination. Soil and Sediment Contamination: An International Journal. 2020;**29**(4):369-383

[18] Ouyang W, Liu H, Murygina V, Yu Y, Xiu Z, Kalyuzhnyi S. Comparison of bio-augmentation and composting for remediation of oily sludge: A field-scale study in China. Process Biochemistry. 2005;**40**(12):3763-3768

[19] Magoha HS. Destruction of polycyclic aromatic hydrocarbons (pah's) and aliphatic hydrocarbons in soil using ball milling[thesis]. Auckland, New Zealand: Auckland University of Technology; 2004

[20] Kaupp G. Mechanochemistry: The varied applications of mechanical bond-breaking. CrystEngComm. 2009;**11**:388-403

[21] Fermor TR. Applied aspects of composting and bioconversion of lignocellulosic materials an overview. International Bio Deterioration and Biodegradation. 1993;**31**:87-106

[22] Tuomela M, Vikman M, Hatakka A, Itävaara M. Biodegradation of lignin in a compost environment: A review. Bioresource Technology. 2000;**72** (**2**:169-183

[23] Filauro G, Andreotti G, Arlotti D, Reisinger HJ. Blow out of Trecate 24 crude oil well: How bioremediation techniques are solving a major environmental emergency in a valuable agricultural area. In: Contaminated Soil 1998. 1998

[24] Jørgensen KS, Puustinen J, Suortti AM. Bioremediation of petroleum hydrocarbon-contaminated soil by composting in biopiles. Environmental Pollution. 2000;**107**(2):245-254

[25] Mills MA, Bonner JS, McDonald TJ, Page CA, Autenrieth RL. Intrinsic bioremediation of a petroleum-impacted wetland. Marine Pollution Bulletin. 2003;**46**(7):887-899

[26] Garbisu C, Alkorta I. Review paper: Phytoextraction: A cost-effective plant-based technology for the removal of metals from the environment. Bioresource Technology. 2001;**77**(3):229-236

[27] Wiltse CC, Rooney WL, Chen Z, Schwab AP, Banks MK. Technical report-greenhouse evaluation of agronomic and crude oil-phytoremediation potential among alfalfa genotypes. Journal of Environmental Quality. 1998;**27**(1):169-173

[28] Zhang W, Zhang S, Wan C, Dapan W, Youbin Y, Wang X. Source diagnostics of polycyclic aromatic hydrocarbons in urban road runoff, dust, rain and canopy throughfall. Environmental Pollution. 2008;**153**:594-601

[29] Pies C, Hoffman B, Petrowsky J, Yang Y, Ternes TA, Hofmann T. Characterization and source identification of polycyclic aromatic hydrocarbons (PAHs) in river bank soils. Chemosphere. 2008;**72**:1594-1601

[30] Yunker MB, Macdonald RW, Vingarzan R, Mitchell RH, Goyette D, Sylvestre S. PAHs in the Fraser River basin: A critical appraisal of PAH ratios as indicators of PAH source and composition. Organic Geochemistry. 2002;**33**:489-515

[31] Igbiri S, Udowelle NA, Ekhator OC, Asomugha RN, Igweze ZN, Oriskwe OE. Polycyclic aromatic hydrocarbons in edible Mushrooms from Niger Delta, Nigeria: Carcinogenic and non-carcinogenic health risk assessment. Asian Pacific Journal of Cancer Prevention. 2017;**18**:437-447

[32] Nisbet CT, LaGoy PK. Toxic equivalency factors (TEFs) for polycyclic aromatic hydrocarbons (PAHs). Regulatory Toxicology and Pharmacology. 1992;**16**(3):290-300

[33] Semple KT, Morriss AWJ, Paton GI. Bioavailability of hydrophobic organic contaminants in soils: Fundamental concepts and techniques for analysis. European Journal of Soil Science. 2003;**54**(4):809-818

[34] Atagana HI. Co-composting of PAH-contaminated soil with poultry manure. Letters in Applied Microbiology. 2004;**39**(2):163-168

[35] Juhasz AL, Naidu R. Review: Bioremediation of high molecular weight polycyclic aromatic hydrocarbons: A review of the microbial degradation of benzo[*a*]pyrene. International Biodeterioration & Biodegradation. 2000;**45**(2):57-88

[36] Shuttleworth KL, Cerniglia E. Environmental aspects of PAH biodegradation. Applied Biochemistry and Biotechnology. 1995;**54**:291-302

[37] Kanaly R, Harayama S. Biodegradation of high- molecular weight polycyclic aromatic hydrocarbons by bacteria. Journal of Bacteriology. 2000;**182**:2059-2067

[38] Chijioke-Osuji CC, Ibegbulam-Njoku PN, Belford EJD. Biodegradation of crude oil polluted soil by Co-composting with agricultural wastes and inorganic fertilizer. Journal of Natural Sciences. 2014;**4**(6):2224-3186

[39] Agarry SE, Owabor CN, Yusuf RO. Bioremediation of soil artificially contaminated with petroleum hydrocarbon oil mixtures: Evaluation of the use of animal manure and chemical fertilizer. Bioremediation Journal. 2010;**14**:189-195

[40] Phan C-W, Sabaratnam V. Potential uses of spent mushroom substrate and its associated lignocellulosic enzymes. Applied Microbiology and Biotechnology. 2012;**96**:863-873

[41] Boyle CD, Kropp BR, Reid ID. Solubilization and mineralization of lignin by white rot fungi. Applied and Environmental Microbiology. 1992;**58**(10):3217-3224

[42] Song H-G. Comparison of pyrene biodegradation by white rot fungi. World Journal of Microbiology and Biotechnology. 1999;**15**:669-672

[43] Ribas LCC, de Mendonça MM, Camelini CM, Soares CHL. Use of spent mushroom substrates from *Agaricus subrufescens* (syn. *A. Blazei*, *A. Brasiliensis*) and Lentinula edodes productions in the enrichment of a soil-based potting media for lettuce (Lactuca sativa) cultivation: Growth promotion and soil bioremediation. Bioresource Technology. 2009;**100**:4750-4757

[44] Li X, Wu Y, Lin X, Zhang J, Zeng J. Dissipation of polycyclic aromatic hydrocarbons (PAHs) in soil microcosms amended with mushroom cultivation substrate. Soil Biology and Biochemistry. 2012;**47**:191-197

[45] Chan KY, Van Zwieten L, Meszaros I, Downie A, Joseph S. Using poultry litter biochars as soil amendments. Australian Journal of Soil Research. 2008;**46**(5):437-444

[46] Paladino G, Arrigoni JP, Satti P, Morelli I, Mora V, Laos F. Bioremediation of heavily hydrocarbon-contaminated drilling wastes by composting. International Journal of Environmental Science and Technology. 2016;**13**:2227-2238

[47] Nam JJ, Thomas GO, Jaward FM, Steinnes E, Gustafsson O, Jones KC. PAHs in background soils from Western Europe: Influence of atmospheric deposition and soil organic matter. Chemosphere. 2008;**70**:1596-1602

[48] Orecchio S. Contamination from polycyclic aromatic hydrocarbons (PAHs) in the soil of a botanic garden localized next to a former manufacturing gas plant in Palermo (Italy). Journal of Hazardous Materials. 2010;**180**:590-601

[49] Zhang J, Yang JC, Wang R, Hou H, Du X, Fan S, et al. Effects of pollution sources and soil properties on the distribution of polycyclic aromatic hydrocarbons and risk assessment. Science of the Total Environment. 2013;**464**:1-10

[50] Zhao X, Ding J, You H. Spatial distribution and temporal trends of polycyclic aromatic hydrocarbons (PAHs) in water and sediment from Songhua River, China. Environmental Geochemistry and Health. 2014;**36**:131-143

[51] Yuan P, Li T, Ding X, Zhao G, Ye S. Distribution, sources and potential toxicological significance of polycyclic aromatic hydrocarbons (PAHs) in surface soils of the Yellow River Delta, China. Marine Pollution Bulletin. 2014;**83**:258-264

[52] CCME (Canadian Council of Ministers of the Environment). Canadian Soil Quality Guidelines for Potentially Carcinogenic and Other PAHs: Scientific Criteria Document. Winnipeg: CCME; 2010

[53] Smith VH, Graham DW, Cleland DD. Application of resource-ratio theory to hydrocarbon biodegradation. Environmental Science & Technology. 1998;**32**:3386-3395

[54] Braddock JF, Ruth ML, Catterall PH. Enhancement and inhibition of microbial activity in hydrocarbon-contaminated arctic soils: Implications for nutrient-amended bioremediation. Environmental Science and Technology. 1997;**31**:2078-2084

Chapter 6

Endophytic Microorganisms: Harnessing Biological Control for Biotic and Abiotic Stress Management in Plants

Soheila Aghaei Dargiri, Shahram Naeimi and Mojtaba Khayam Nekouei

Abstract

Strengthening agriculture is crucial for ensuring food security for the world's expanding population. Endophytes, microorganisms that live within plant tissues without harming the host, can colonize these tissues. They promote plant growth, serve as biocontrol agents, and naturally shield the host from a variety of biotic and abiotic challenges. Each year, the emergence of numerous stresses in crops leads to a decrease in yield, resulting in significant losses. To mitigate these losses and manage plant pests and diseases, various effective strategies should be employed. Endophytes, being environmentally friendly, non-toxic, and cost-effective, could serve as an alternative for farmers aiming for sustainable agriculture. This chapter will discuss the biochemical, molecular, and genetic mechanisms of endophytes in enhancing stress tolerance in different crops and their role as biological control agents. Furthermore, endophytes and their metabolites should be explored as potential beneficial agents in the biological control of plant pests and diseases, extending the investigation up to the multi-omics level.

Keywords: bacterial endophytes, fungal endophytes, biocontrol, abiotic stress, biotic stress, phytopathogen

1. Introduction

Sustainable agriculture aims to address climate change, a major hurdle in achieving food security for the expanding global population. This predicts a 1.7–4.8°C rise in global temperatures by the twenty-first century, altering precipitation patterns [1, 2]. Over the past three decades, climatic changes have induced abiotic stress, reducing global agricultural productivity by 1–5% and necessitating a 70–100% increase in food supply by 2050 [3]. Climate change poses a significant threat to ecology and impacts both short-term and long-term food production, primarily through greenhouse gas (GHG) emissions [3].

The excessive use of synthetic pesticides during the green revolution has predicted adverse environmental impacts, including pesticide resistance, insect resurgence, harm to nontarget organisms, and environmental pollution. This necessitates a long-term conservation strategy to protect plants [4, 5]. A comprehensive solution is needed to eliminate the insecurities associated with food production and environmental conservation, which are exacerbated by various pests and diseases that infest crops, such as fungi, bacteria, viruses, and nematodes. Alternative strategies that protect plants using microorganisms and biological control agents appear more reliable and environmentally friendly [6]. Key microorganisms associated with plants, such as endophytes, could play a vital role in mitigating crop responses to climate change [7, 8].

Endophytic microorganisms, which live within plant tissues without causing harm, are the subject of extensive global research aimed at enhancing agricultural and environmental sustainability [9]. These organisms are typically fungi or bacteria [10]. Studies suggest that endophytes offer a promising solution for managing pests, plant diseases, and environmental stresses [11–13]. Therefore, the use of naturally occurring beneficial endophytes for disease management could be an alternate approach to improving the resilience of crop varieties and plant resistance [14]. This would reduce chemical inputs and environmental stress without negative impacts. Beneficial endophytic bacteria present in plant tissues provide a a potential natural solution for stressed environments [15]. Upon inoculation, endophytic microorganisms are known to modify their hosts' genome, epigenome, proteome, and metabolome to adapt to biotic and abiotic stressors. Plants inoculated with endophytes develop a modified "ome," enhancing their ability to withstand various biotic and abiotic challenges, such as diseases, salt, and drought. The molecular mechanisms through which endophytes aid crops in coping with abiotic stress remain unclear. However, recent advances in mass spectroscopy (MS) and high-throughput sequencing technology (HTST) have raised hopes for a comprehensive examination of genes and proteins that will elucidate the molecular interactions between plant endophytes under both abiotic and abiotic stress conditions [16, 17]. This chapter discusses the biochemical and genetic pathways of endophytic microorganisms as biological control agents in the managing biotic and abiotic stresses.

2. Abiotic and biotic stresses in plants

Plants can be affected by various pressures, including biotic stress inducers, nutritional or vital ion shortages, or antagonistic or poisonous compounds in their surroundings [18]. Biotic and abiotic stressors account for 30–50% of agricultural losses worldwide [19]. The main abiotic stressors are temperature, salinity, drought, flooding, trace metals, and nutritional deficiencies (**Figure 1**). Biotic stressors can be caused by pathogenic microbes up to insects or nematodes.

3. Endophytic microorganisms

Endophytic microorganisms, such as bacteria, fungus, and actinomycetes, are prevalent in the tissues of all plant species and do not harm or cause disease [9]. Numerous microbial communities are associated with plants, and these communities can respond to their host plant in a positive, negative, or neutral way. The subgroup of

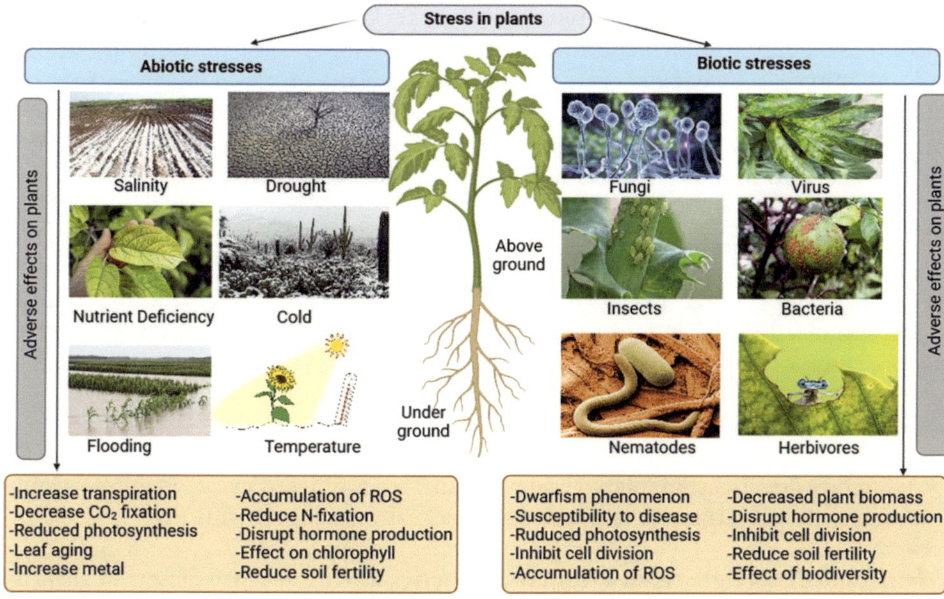

Figure 1.
Effect of abiotic and biotic stress on plants (from the authors' study).

the rhizosphere microbiome known as plant growth-promoting endophytes significantly influences plant microecosystems [20, 21]. They coexist and multiply in plant tissues, such as stems, roots, seeds, fruit, buds, and leaves, without endangering the host plants [22].

3.1 Endophyte diversity and classification

Endophytes or internal microorganisms are a group of microbes and fungi that live inside plant tissues and organs. These microbes are usually in a symbiotic relationship with plants and have multiple effects on plant health and performance [23]. Below are some types of endophytes, their classification and distribution in plants and plant environments:

1. Endophytic fungi: this includes fungi such as *Sclerotinia*, *Fusarium*, *Trichoderma*, and *Penicillium*. These fungi can form the roots, stems, and leaves of living plants [24].

2. Endophytic bacteria: bacteria, such as rhizobium, rhizovactinum, endophytes, and other endophytes can live in plant tissues and provide nitrogen fixation, growth hormone production, and stress resistance [25].

3. Endophytic viruses: these are viruses that live inside plants and can affect the structure and function of the plant [26].

The classification of these endophytes is usually based on their physiological, morphological, and genetic characteristics. Also, they can be classified based on host types (plant species), environmental requirements, and biological effects [27].

The distribution of these endophytes also varies with the type of plant, different environmental conditions, including soil, amount, temperature, and amount of light. Some endophytes may coexist exclusively with one plant species, while they may be found in different plant species [28]. Consequently, further studies in this field can contribute to a better understanding of the types of endophytes, their role in plant maintenance, and their contribution to plant health and performance [29].

3.2 Methods for studying endophytes

Studying endophytes involves various methods and techniques to isolate, identify, and understand their interactions with host plants and their environments [30]. Here are some common methods used in the study of endophytes:

1. Isolation and cultivation:

    - Surface sterilization: plant tissues are surface sterilized using chemical agents (e.g., bleach, ethanol) to remove superficial microorganisms [31].

    - Tissue dissection: tissues (e.g., roots, stems, and leaves) are dissected aseptically to access endophytes residing within plant tissues [32].

    - Culture media: endophytes are cultured on specific growth media optimized for their growth and isolation [33].

    - Incubation conditions: cultures are incubated under controlled conditions (e.g., temperature, humidity) to promote the growth of endophytic microorganisms [34, 35].

2. Molecular techniques:

    - DNA extraction: DNA is extracted from isolated endophytes using various extraction methods [36].

    - PCR (polymerase chain reaction): PCR techniques are employed to amplify specific regions of microbial DNA (e.g., 16S ribosomal RNA (rRNA) for bacteria and internal transcribed spacer (ITS) region for fungi) for identification and phylogenetic analysis [37].

    - Sequencing: DNA sequencing technologies (e.g., Sanger sequencing, Next-Generation Sequencing) are utilized to determine the genetic sequences of endophytes for species identification and diversity analysis [38].

    - Metagenomics and metatranscriptomics: these approaches analyze the collective genetic material (DNA or RNA) of microbial communities, including endophytes, within plant tissues without the need for cultivation [39].

3. Microscopic examination

    - Light microscopy: microscopic examination of stained tissue sections or microbial cultures allows for the visualization and morphological characterization of endophytes [40].

- Fluorescence microscopy: fluorescent dyes can be used to visualize endophytes within plant tissues, providing insights into their localization and interactions with host cells [41].

4. Physiological and functional studies

- Physiological assays: Biochemical and physiological assays are conducted to assess the metabolic activities and functional traits of endophytes, such as nitrogen fixation, hormone production, and disease suppression [42].
- Inoculation experiments: Inoculation of plants with purified endophyte cultures or community mixtures helps to elucidate the effects of endophytes on host growth, stress tolerance, and disease resistance [43].

5. Ecological and population studies:

- Field surveys: Collection and analysis of plant and soil samples from natural habitats provide insights into the diversity, distribution, and ecological roles of endophytes in different ecosystems [44].
- Longitudinal studies: Monitoring changes in endophyte communities over time in response to environmental factors, host plant development stages, and disturbances (e.g., climate change and land use change) [45].

6. Bioinformatics and data analysis:

- Bioinformatics tools and software packages are used to analyze sequencing data, construct phylogenetic trees, and assess microbial diversity and community composition [46].
- Statistical analyses help to interpret experimental results, identify significant associations between endophytes and host plants, and infer ecological patterns and processes [47].

By integrating these methods, researchers can gain a comprehensive understanding of the diversity, ecology, and functions of endophytes in plant-microbe interactions.

3.3 Case studies of endophyte application in agriculture

Case studies and applications of endophytes in agriculture are currently one of the most used and attractive fields in agricultural science research. Endophytes, as internal microorganisms of plants, can have positive effects on the growth, development, and performance of plants. Below are some case studies of endophyte applications in agriculture:

3.3.1 Increasing resistance to environmental stresses

Studies have shown that some endophytes can provide plants with greater resistance to environmental stresses, such as drought, salinity, temperature, and plant diseases [16]. For example, some endophytes can improve plant resistance to soil salinity and promote root system development under stressful conditions [48].

3.3.2 Nitrogen fixation

Endophytic bacteria such as rhizobium are able to convert molecular nitrogen into forms that can be absorbed by the plant and stabilize it in the soil by creating symbiotic relationships with self-feeding plants. This application provides improvement of soil quality and nitrogen efficiency in agriculture [49].

3.3.3 Increasing the growth and yield of plants

Some endophytes can increase plant performance. This increase may be due to increased uptake of nutrients, production of plant growth hormones, and control of diseases and pests [50].

3.3.4 Biological control of pests and diseases

Some endophytes can act as biological control agents to control plant pests and diseases. This method reduces the need to use chemical poisons in agriculture, and on the other hand, it also serves the environment and human health [51].

3.3.5 Improving the quality of agricultural products

Endophytes can make significant improvements in the quality of agricultural products, such as increasing nutrient content, producing antioxidants, and improving the appearance of products [52].

In general, the use of endophytes in agriculture is considered as a sustainable and effective way to increase productivity and preserve the environment, and further research in this field continues.

3.4 Role of endophyte in abiotic stress management

Endophytes govern multiple host activities such as immune system stimulation, growth and development, and resistance to biotic and abiotic challenges in symbiotic partnerships with a wide range of plant species [9]. They primarily provide stress resistance on plants by (1) triggering the host plant's stress-regulating system [53, 54] and (2) producing antistress chemicals on their own [55, 56]. Abiotic stresses, such as heat, salinity, and drought, impact plants and cause cellular and molecular deterioration by lowering productivity and yield [57, 58]. In response to these abiotic stresses, endophytes produce scavenger molecules like reactive oxygen species (ROS), upregulate certain stress-responsive genes, synthesize antistress metabolites, and promote the accumulation of certain suitable solutes in plants [59]. Plants can withstand abiotic stress due to phytohormones generated by endophytes, including salicylic acid, jasmonic acid, gibberellic acid, indole acetic acid (IAA), and brassinosteroids [60, 61].

3.5 Role of endophyte in biotic stress management

Biotic stress, particularly that brought on by bacterial and fungal phytopathogens, is the primary source of pre- and post-harvest losses in the agricultural sector [62].

A plant's generalized defense mechanism cannot completely release pressure or satisfy the requirements of multitrees tolerance for a plant to grow and survive. Current strategies for obtaining stress-resistant cultivars include genetic engineering, as well as chemical and physical techniques. However, their ability to withstand stress is short-lived, and they are not environmentally sustainable. Thus, utilizing naturally occurring beneficial endophytes for disease management could be an alternate method for enhancing crop types' resilience and plant resistance [14]. This will lessen environmental stress without having a negative impact, in addition to lowering chemical inputs. Beneficial endophytic bacteria found in plant tissues offer a possible natural remedy for stressed environments. Endophytes, which are fungi or bacteria, have enormous potential for use as biocontrol agents. They oppose disease-causing phytopathogens and also lessen the damage that the pathogens are assumed to have caused. They generate different antioxidants and a number of functional antiviral and antibacterial metabolites to inhibit infections [63].

4. Biological control

Plant diseases pose a constant threat to global food security. Existing tools often fall short of controlling these diseases and minimizing losses. These diseases harm food and plant resources and limit output in both tropical and temperate regions of the world. Despite their evolutionary differences (an oomycete and an invertebrate species, respectively), there are some management-related commonalities between them. The wide variety of hosts combined with the emergence of virulent strains/populations makes it challenging to use nonhost crops and resistant cultivars [64]. The use of chemical pesticides is not always practical or efficient due to insensitive strains, cost, or application technique [65]. Moreover, the pesticide industry has faced challenges in developing new pesticide compounds. Society expects higher quality and less environmental impact in addition to food production security [66]. These factors have driven the search for environmentally friendly, innovative, and efficient pest management strategies, making biopesticides a valuable tool in combating plant disease losses [67].

So, what exactly is biological control? The traditional definition of biological control is a reduction in a pathogen population (inoculum) or in the disease determinants caused by an organism other than a human or plant [68]. It is also described as an effort to use preexisting relationships and natural processes to transfer a common event from nature to agricultural systems [69]. However, most, if not all, biological control agents have proven to be capable of intimately interacting with and/or colonizing plants in some way. They have established a sophisticated system of interkingdom communication with plants that uses a molecular language for signaling [70].

4.1 Biocontrol mechanisms under bacterial endophytes

Bacterial endophytes that promote plant growth can help limit the growth of plant diseases. They trigger induced systemic resistance (ISR), an innate plant defense mechanism, and they also synthesize a range of antibiotic chemicals and enzymes with antagonistic action toward phytopathogens [71]. The ability to invade plant tissues may be another biocontrol mechanism, although quantitative research on this is challenging (even in dwellers of the rhizosphere) [72]. However, there is some evidence to suggest that the colonization and competition of beneficial bacteria can reduce the severity and frequency of plant diseases (**Figure 2**) [73].

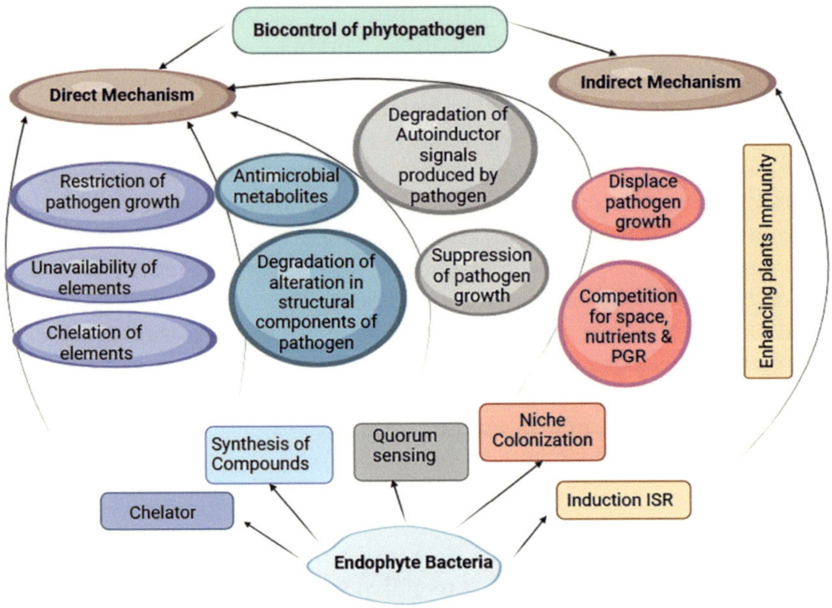

Figure 2.
Mechanism of endophytic bacteria in biocontrol (from to the authors' study).

4.2 Biocontrol mechanisms under fungal endophytes

In general, a few key concepts are accepted for the study of biological control agents (BCAs) and endophytic fungi. These include the following: (i) endophyte-induced activation of the plant defense system; (ii) suppression through mycoparasitism; (iii) inhibition through antibiosis; and (iv) competition for nutrients and space [74].

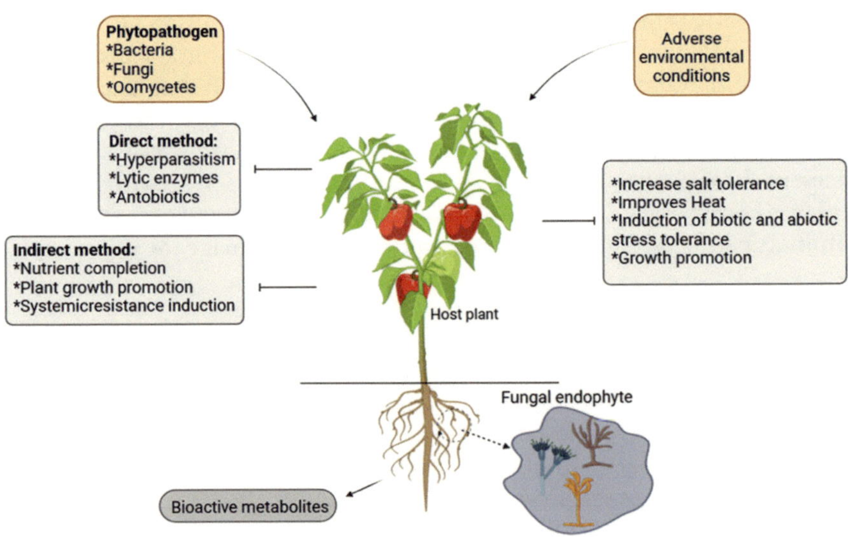

Figure 3.
Mechanism of endophytic fungal in biocontrol (from the authors study).

Additionally, multiple processes may be triggered simultaneously and at most times. In general, improving plant health and suppressing disease can be achieved by modifying the level of plant growth hormone or acquiring nutrients to promote plant growth [75]. Endophytes can serve as a biocontrol agent in addition to providing protection to the host plant throughout its life cycle. To prevent plant diseases and promote sustainable agriculture, endophytes are used as biocontrol agents. Endophytes can produce metabolites that are antibacterial, insecticidal, antioxidant, antitumor, and antiviral, in addition to having antagonistic activity against phytopathogens that cause disease (**Figure 3**).

5. Endophytic microbes and pollution remediation

Endophytes can reduce disease and protect plants through different ways:

1. Competition and antagonism: endophytes compete with pathogens for space and nutrients within the plant [76]. By occupying niches that pathogens would otherwise exploit, endophytes can effectively inhibit the growth and establishment of pathogens [77].

2. Production of antimicrobial compounds: many endophytes produce secondary metabolites with antimicrobial properties [78]. These compounds can directly inhibit the growth of plant pathogens, providing a protective shield for the host plant [79].

3. Induced systemic resistance (ISR): endophytes can trigger the plant's defense mechanisms, enhancing its ability to resist diseases [80]. This is often achieved through the activation of systemic acquired resistance (SAR) or induced systemic resistance (ISR), which primes the plant's immune system for a more effective response against pathogens [81].

4. Enhancement of plant fitness: some endophytes promote overall plant health and fitness, making the plant less susceptible to diseases [82]. This can involve various mechanisms, such as improving nutrient uptake, enhancing stress tolerance, and stimulating growth [83].

5. Biocontrol agents: certain endophytes act as biocontrol agents, actively attacking and killing pathogens within the plant tissues. This direct antagonism can significantly reduce disease severity and incidence [84].

6. Cross-protection: endophytes may confer resistance to multiple pathogens simultaneously or provide protection against different types of stresses. This cross-protection can be particularly valuable in environments where plants are exposed to multiple disease threats [85].

Overall, harnessing the potential of endophytes holds promise for sustainable disease management in agriculture and forestry. By exploiting the natural symbiotic relationship between plants and these beneficial microorganisms, it's possible to develop eco-friendly strategies for disease control while reducing reliance on chemical pesticides.

6. The potential application of endophytes in phytopathogen management

Microbial biocontrol agents have been widely used recently to manage pests and diseases when crops are being stored pre- or post-harvest [86]. Microbial antagonists are popular worldwide due to their nontoxic and affordable nature. In addition to negatively affecting the natural texture and quality of crops and fruits, readily available chemical pesticides are also harmful to consumers' health. Farmers and horticulturists extensively use these pesticides in the field and during storage. In the past 20 years, there has been an increase in the use of microbial antagonistic pest and disease management [86]. However, because endophytic bacteria may better colonize and acclimatize to stressful environments than other microbial species, using them as biocontrol agents has greater advantages [87]. All of these rhizospheric microorganisms—mycoparasitism, competition for nutrients and space, lytic enzyme production, bioactive compound synthesis, antibiotics, antimicrobials, and volatiles—have biocontrol mechanisms that directly impede the growth of phytopathogens and reduce the incidence of disease. These mechanisms are also shared by endophytes. They protect against pathogen invasion while subtly activating the host plants' systemic and local plant defense mechanisms [88].

7. Plant stress tolerance mediated by endophytes

Potential endophyte entry, colonization, and interactions with host plant tissues are summarized in **Figure 4**. Plant pattern recognition receptors (PRRs) blend with the chitin deacetylase enzymes that fungal endophytes produce to enter the host plant [89]. According to additional data, certain endophytic bacteria may release their microbe-associated molecular patterns (MAMPs), which could cause plant PRRs to misidentify the bacterium or trigger a weak and temporary defense response [90].

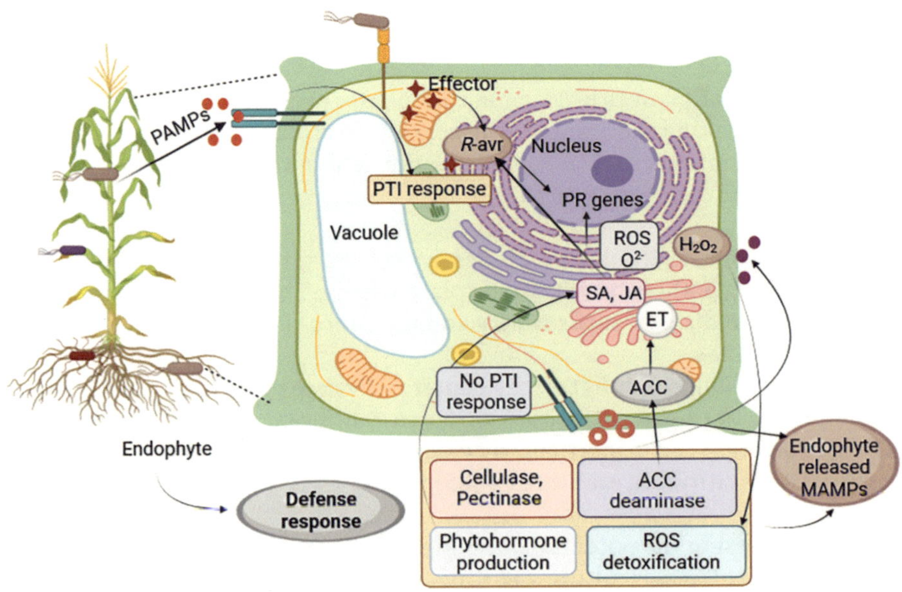

Figure 4.
Plants and endophytes interaction. The figure depicts the interaction between endophytes and their hosts. It highlights how the type of endophyte influences their colonization of a plant and their interaction with it.

Bacterial endophytes often enter through apertures created by the elongation zones or the formation of lateral roots [91]. On the other hand, fungal endophytes can enter the plant as spores through herbivory or insects, or they can spread vertically, from mother plants to the seeds of their progeny plants [92]. To facilitate their entry, certain endophytes also produce enzymes that break down cell walls. These endophytes are not recognized by the host plant's PRRs due to their unique MAMPs. Endophytes assist plants in responding to stress by producing phytohormones and detoxifying the reactive oxygen species (ROS) [93]. The figure also includes a key to abbreviations such as JA is for jasmonic acid, SA stands for salicylic acid, PAMPs for pathogen-associated molecular patterns, PTI for pathogen-triggered immunity, ACC for aminocyclopropane-1-carboxylic acid, ROS for reactive oxygen species, MAMPs for microbe-associated molecular patterns, PR for pathogenesis-related, and ET for ethylene. R-avr is a virulence effector, or resistance gene. This information is part of the author's study [94].

8. Biochemical and genetic pathways of endophytes in tolerance to biotic and abiotic stresses

8.1 Biochemical and genetic pathways of endophytes in abiotic stress

Endophytic bacteria and fungi play a crucial role in enhancing plant resistance to abiotic stressors through genetic and metabolic mechanisms [6, 95–97]. The genetic routes involve the interaction of endophytes with the host plant's genome, which can trigger plant resistance through gene expression [6, 95]. The metabolic routes involve the production of phytohormones and the detoxification of reactive oxygen species (ROS) by endophytes, which help plants respond to stress [6, 96, 97]. The following provides a thorough explanation of the genetic and metabolic routes used by endophytes under abiotic stress.

8.1.1 Biochemical pathways

8.1.1.1 Antioxidant enzymes

Endophytic plants produce antioxidant enzymes, including peroxidase (POD), catalase (CAT), and superoxide dismutase (SOD). These enzymes assist in lessening oxidative damage to plant cells by scavenging reactive oxygen species (ROS) produced under abiotic stress [98].

8.1.1.2 Osmoprotectant accumulation

Endophytes aid in the production and accumulation of osmoprotectants like proline, glycine betaine, and soluble carbohydrates in plant cells. In situations of water shortage or salinity, these substances function as osmolytes, assisting plants in maintaining cellular turgor and safeguarding cellular structures [99].

8.1.1.3 Abscisic acid (ABA) regulation

Abscisic acid, a crucial plant hormone involved in stress responses, is regulated by endophytes. Under drought stress, ABA promotes the production of stress-responsive genes and aids in stomatal closure, and reduces water loss [100].

8.1.1.4 Induction of heat shock proteins (HSPs)

Heat shock proteins are molecular chaperones that are induced to express by endophytes. Under conditions of heat stress, HSPs help fold proteins, prevent protein denaturation, and protect cellular proteins overall [101].

8.1.1.5 Phytohormone crosstalk

Endophytes control the relative amounts of phytohormones, such as ethylene (ET), salicylic acid (SA), and jasmonic acid (JA). The interactions between these hormones significantly influence plant responses to a variety of abiotic stressors [102].

8.1.2 Genetic pathways

8.1.2.1 Activation of stress-responsive genes

When endophytes are present, plant hosts' stress-responsive genes are made to express more. These genes could produce proteins that improve stress tolerance through different mechanisms such as antioxidant defense and osmoprotectant [99].

8.1.2.2 Transcription factor regulation

Endophytes affect the transcription factors responsible for controlling the expression of genes that respond to stress. For example, dehydration-responsive element-binding/C-repeat binding factor (DREB/CBF) transcription factors promote tolerance to cold stress [103, 104].

8.1.2.3 Epigenetic modifications

Endophytes have the ability to cause epigenetic changes that control the expression of genes linked to stress, such as histone acetylation and DNA methylation. These adjustments result in long-term changes in the plant's stress response [105].

8.1.2.4 RNA-mediated regulation

Endophytes possess the remarkable ability to modulate RNA-based regulatory systems within plants. These regulatory pathways involve small interfering RNA (siRNA) and microRNA (miRNA). The impact of these compounds extends to post-transcriptional gene silencing and the control of abiotic stress response in plants [106]. Furthermore, endophytes establish symbiotic connections with plants, significantly contributing to the host's defense against biological stressors. This defense mechanism operates through intricate biochemical and genetic pathways [107]. Notably, the specific mechanisms employed by endophytes may vary depending on the type of endophyte and the specific of biotic stress faced [80].

8.2 Biochemical and genetic pathways of endophytes in biotic stress

The interaction between endophytes and plants involves complex genetic and metabolic processes. These mechanisms enhance the host's resistance to biotic stresses [93]. Let us delve into a detailed exploration of these pathways:

8.2.1 Biochemical pathways

8.2.1.1 Antibiotic and antifungal compound production

Endophytes play a crucial role in inhibiting the growth and development of pathogenic microorganisms. They achieve this through the production of antibiotics and antifungal chemicals. These bioactive substances serve as direct defense mechanisms against biotic stress [108].

8.2.1.2 Induced systemic resistance (ISR)

Endophytes induce systemic resistance in plants by releasing signaling molecules, such as ethylene (ET), salicylic acid (SA), and jasmonic acid (JA) [81]. These signaling pathways activate the host plant's defense mechanisms, fortifying its ability to fend off future pathogen attacks [109].

8.2.1.3 Cell wall-degrading enzymes

Endophytes produce specialized enzymes, including gluconates and chitinases, which play a crucial role in breaking down pathogen cell walls. By doing so, endophytes hinder pathogen invasion and protect plant tissues [84].

8.2.1.4 Phytoalexin accumulation

Phytoalexins are bioactive molecules synthesized by plants as a defense mechanism against pathogens. Endophytes actively promote the accumulation of phytoalexins, bolstering the ability to combat biotic stress [110].

8.2.1.5 Modulation of plant hormones

Endophytes exert influence over plant growth and development by regulating plant hormones such as auxins, cytokinins, and gibberellins. This hormonal modulation indirectly contributes to the plant's resistance against biotic stress [111].

8.2.2 Genetic pathways

8.2.2.1 Activation of pathogenesis-related (PR) proteins

Endophytes express pathogenesis-related (PR) proteins within the host plant. These proteins possess antibacterial properties and are pivotal for the plant's defense against infections [112].

8.2.2.2 Transcriptional regulation of defense genes

Endophytes modulate the expression of plant defense genes. Among these genes are those encoding antimicrobial peptides, secondary metabolite-producing enzymes, and other defense-related proteins, all of which experience upregulation [113].

8.2.2.3 Recognition and signaling pathways

Endophytes engage in intricate recognition and signaling pathways during interaction with plants. Plant receptors detect microbial patterns, triggering subsequent signaling cascades that fortify the plant's defense mechanisms [93].

8.2.2.4 MicroRNA (miRNA) and small interfering RNA (siRNA) pathways

Endophytes exert influence over RNA-based regulatory pathways, including miRNA and siRNA. These pathways play pivotal roles in post-transcriptional gene silencing and modulate a plant's response to external stressors [2].

8.2.2.5 Epigenetic modifications

Endophytes induce epigenetic alterations within plants, such as histone acetylation and DNA methylation. These modifications regulate the expression of defense-related genes, leading to dynamic changes in the plant's response to biotic stress over time [114]. Through their cooperative engagement in these intricate biochemical and genetic pathways, endophytes establish a symbiotic connection with plants, bolstering the host's resistance against biotic stressors [30]. Notably, the specific mechanisms employed by endophytes may vary based on the type of endophyte and the specific biotic stress encountered [80].

9. Challenges and limitations in endophyte research and application

The use of endophytes in agriculture and other fields of application faces challenges and limitations, which include the following:

1. Identification and isolation: identification and isolation of endophytes is often time-consuming and difficult. Some endophytes may not be suitable for growth in laboratory conditions and require special conditions, which complicates the process of isolation and identification [115].

2. Stability and biogenetics: the stability of endophytes in different environments and conditions can be a challenge. Many endophytes survive only in certain conditions of the host and lose their ability to grow and function in conditions outside their natural environment [116].

3. Effective use: effective use of endophytes means providing efficient and practical methods for farmers. This includes efficient isolation, preservation, and application processes for farms and agricultural environments [117].

4. Side effects: side and unwanted effects of endophytes may cause problems for plants or the environment. For example, some endophytes may unexpectedly have deleterious effects on plant growth and development or increase resistance to agrochemicals [118].

5. Legal issues and regulations: the use of endophytes in agriculture may be accompanied by legal regulations and restrictions related to agriculture and

the environment. For the use of endophytes as pest and disease control agents, there can be strict regulations that require case-by-case approvals and special permits [119].

Nevertheless, considering the great potential that endophytes have for improving agricultural systems, continuous efforts are being made to manage the existing challenges and limitations and provide effective solutions for the optimal use of these microorganisms.

10. Future and perspective

Endophytic microorganisms hold significant promise for sustainable agriculture and ecosystem resilience when employed as biological control agents to manage both biotic and abiotic stresses. As we look ahead, several exciting developments are on the horizon:

- Advancing understanding of endophyte-plant interactions: ongoing progress in microbial ecology and genetics will deepen our comprehension of the intricate interactions between endophytes and plants. This knowledge will pave the way for the development of specialized and targeted therapies.

- Harnessing modern technologies:

    - Metagenomics: leveraging metagenomic approaches will allow us to explore the genetic diversity of endophytes comprehensively. This knowledge can inform strategies to design or enhance advantageous traits in these microorganisms.

    - Synthetic biology: by applying synthetic biology techniques, we can engineer endophytes with specific stress-reducing capabilities. These modifications will enhance their effectiveness in mitigating plant stress.

- Endophytic consortia for robust defense:

    - Combining endophytes with diverse functional strengths into consortia may yield a comprehensive defense against a wide spectrum of biotic and abiotic stresses. Synergistic interactions within these consortia can amplify their protective effects.

- Precision agriculture and customized endophyte selection:

    - Crop stress profiles: precision agriculture will enable tailored endophyte selection based on specific crop stress profiles. Matching endophytes to the unique needs of each crop will optimize stress management.

    - Environmental factors: considering environmental conditions (such as soil type, climate, and local microbiome) will further refine endophyte application strategies.

- A new era of sustainable stress management:
    - As we integrate these advancements, we anticipate a paradigm shift in stress management techniques. Customized endophytes, informed by cutting-edge science, will play a pivotal role in sustainable and highly effective stress mitigation.

11. Conclusion

In the pursuit of environmentally friendly and sustainable crop protection, endophytes emerge as powerful biological control agents, replacing traditional chemical interventions. These remarkable microorganisms not only manage stress but also influence hormone balance and nutrient uptake, thereby enhancing overall plant function.

Here are key takeaways:

- Stress-tolerant endophytes for enhanced crop productivity:
    - Endophytes exhibiting stress tolerance hold immense commercial potential. By harnessing their features, we can develop bioinoculants that significantly improve crop productivity, especially in challenging conditions.
    - These bioinoculants act as allies, bolstering plant resilience against both biotic and abiotic stressors.
- Promoting biodiversity and ecological balance:
    - Leveraging the diversity of endophytes contributes to biodiversity within the plant microbiome. This intricate web of interactions fosters ecological balance and strengthens the overall ecosystem.
- Unleashing the full potential of endophytes:
    - To fully realize the promise of endophytes, continued research, rigorous field testing, and exploration of diverse endophyte strains are essential.
    - Modern techniques, including metagenomics and synthetic biology, will guide the development of specialized endophytes tailored to specific stress scenarios.

In summary, the symbiotic relationships between endophytic microbes and plants offer a promising avenue for addressing the multifaceted challenges posed by agricultural stress. As we delve deeper into this fascinating field, resilient and sustainable agricultural systems will thrive.

Author details

Soheila Aghaei Dargiri[1], Shahram Naeimi[1*] and Mojtaba Khayam Nekouei[2]

1 Department of Biological Control Research, Iranian Research Institute of Plant Protection, Agricultural Research, Education and Extension Organization (AREEO), Tehran, Iran

2 Department of Bioscience, University of Tarbiat Modares, Tehran, Iran

*Address all correspondence to: sh.naeimi@areeo.ac.ir

References

[1] Change IC. Mitigation of climate change. In: Contribution of Working Group III to the Fifth Assessment Report of the Intergovernmental Panel on Climate Change. Vol. 1454. Intergovernmental Panel on Climate Change (IPCC); 2014. p. 147

[2] Zhang R, Zheng F, Wei S, Zhang S, Li G, Cao P, et al. Evolution of disease defense genes and their regulators in plants. International Journal of Molecular Sciences. 2019;**20**(2):335

[3] Pais IP, Reboredo FH, Ramalho JC, Pessoa MF, Lidon FC, Silva MM. Potential impacts of climate change on agriculture-a review. Emirates Journal of Food and Agriculture. 2020;**32**(13):397-407

[4] Meena RS, Kumar S, Datta R, Lal R, Vijayakumar V, Brtnicky M, et al. Impact of agrochemicals on soil microbiota and management: A review. Land. 2020;**9**(2):34

[5] Jacquet F, Jeuffroy M-H, Jouan J, Le Cadre E, Litrico I, Malausa T, et al. Pesticide-free agriculture as a new paradigm for research. Agronomy for Sustainable Development. 2022;**42**(1):8

[6] Adeleke BS, Ayilara MS, Akinola SA, Babalola OO. Biocontrol mechanisms of endophytic fungi. Egyptian Journal of Biological Pest Control. 2022;**32**(1):1-17

[7] Yan L, Zhu J, Zhao X, Shi J, Jiang C, Shao D. Beneficial effects of endophytic fungi colonization on plants. Applied Microbiology and Biotechnology. 2019;**103**:3327-3340

[8] Grabka R, d'Entremont TW, Adams SJ, Walker AK, Tanney JB, Abbasi PA, et al. Fungal endophytes and their role in agricultural plant protection against pests and pathogens. Plants. 2022;**11**(3):384

[9] Anand U, Pal T, Yadav N, Singh VK, Tripathi V, Choudhary KK, et al. Current scenario and future prospects of endophytic microbes: Promising candidates for abiotic and biotic stress management for agricultural and environmental sustainability. Microbial Ecology. 2023;**86**:1-32

[10] Bahram M, Netherway T, Frioux C, Ferretti P, Coelho LP, Geisen S, et al. Metagenomic assessment of the global diversity and distribution of bacteria and fungi. Environmental Microbiology. 2021;**23**(1):316-326

[11] Marasco R, Rolli E, Ettoumi B, Vigani G, Mapelli F, Borin S, et al. A drought resistance-promoting microbiome is selected by root system under desert farming. PLoS One. 2012;**7**(10):e48479

[12] Suman A, Yadav AN, Verma P. Endophytic microbes in crops: Diversity and beneficial impact for sustainable agriculture. In: Microbial Inoculants in Sustainable Agricultural Productivity, Research Perspectives. Vol. 1. 2016. pp. 117-143

[13] Sena L, Mica E, Valè G, Vaccino P, Pecchioni N. Exploring the potential of endophyte-plant interactions for improving crop sustainable yields in a changing climate. Frontiers in Plant Science. 2024;**15**:1349401

[14] Zheng R, Li S, Zhang X, Zhao C. Biological activities of some new secondary metabolites isolated from endophytic fungi: A review study. International Journal of Molecular Sciences. 2021;**22**(2):959

[15] Negi R, Sharma B, Kumar S, Chaubey KK, Kaur T, Devi R, et al. Plant endophytes: Unveiling hidden applications toward agro-environment sustainability. Folia Microbiologica. 2024;**69**(1):181-206

[16] Verma H, Kumar D, Kumar V, Kumari M, Singh SK, Sharma VK, et al. The potential application of endophytes in management of stress from drought and salinity in crop plants. Microorganisms. 2021;**9**(8):1729

[17] Verma A, Shameem N, Jatav HS, Sathyanarayana E, Parray JA, Poczai P, et al. Fungal endophytes to combat biotic and abiotic stresses for climate-smart and sustainable agriculture. Frontiers in Plant Science. 2022;**13**:953836

[18] Khoshru B, Mitra D, Joshi K, Adhikari P, Rion MSI, Fadiji AE, et al. Decrypting the multi-functional biological activators and inducers of defense responses against biotic stresses in plants. Heliyon. 2023;**9**(3):1-26

[19] Akhilesh Kumar AK, Verma J. Does plant-microbe interaction confer stress tolerance in plants: A review? Microbiological Research. 2018;**207**:41-52

[20] Khati P, Parul BP, Nisha KR, Sharma A. Effect of nanozeolite and plant growth promoting rhizobacteria on maize. 3 Biotech. 2018;**8**:1-12

[21] Chaudhary A, Chaudhary P, Upadhyay A, Kumar A, Singh A. Effect of gypsum on plant growth promoting rhizobacteria. Environment and Ecology. 2021;**39**:1248-1256

[22] Chaudhary D, Kumar V, Sharma R, Devi S, Sharma P, Kumar P. Isolation, characterization and management of endophyte(s) prevailing during the in vitro establishment of endangered medicinal herb Himalayan trillium (*Trillium govanianum* Wall. ex D. Don). Plant Cell, Tissue and Organ Culture (PCTOC). 2024;**156**(1):16

[23] Basit A, Shah ST, Ullah I, Ullah I, Mohamed HI. Microbial bioactive compounds produced by endophytes (bacteria and fungi) and their uses in plant health. In: Plant Growth-Promoting Microbes for Sustainable Biotic and Abiotic Stress Management. Cham: Springer; 2021. pp. 285-318

[24] Schouten A. Endophytic Fungi: Definitions, Diversity, Distribution and their Significance in Plant Life. Endophyte Biotechnology: Potential for Agriculture and Pharmacology. The Netherlands: CABI Wallingford UK; Wageningen University & Research Wageningen. 2019. pp. 6-31

[25] Karabulut F, Parray JA, Shafi N, Ikram M. Endophytes: Role in maintaining plant health under stress conditions. In: Plant Endophytes and Secondary Metabolites. Elsevier, Academic Press; 2024. pp. 105-132

[26] Dastogeer KMG, Chakraborty A, Sarker MSA, Akter MA. Roles of fungal endophytes and viruses in mediating drought stress tolerance in plants. International Journal of Agriculture and Biology. 2020;**24**(6):1497-1512

[27] Aamir M, Rai KK, Zehra A, Kumar S, Yadav M, Shukla V, et al. Fungal endophytes: Classification, diversity, ecological role, and their relevance in sustainable agriculture. In: Microbial Endophytes. Woodhead Publishing Series in Food Science, Technology and Nutrition. Elsevier; 2020. pp. 291-323

[28] Ameen M, Mahmood A, Sahkoor A, Zia MA, Ullah MS. The role of endophytes to combat abiotic stress in plants. Plant Stress. 2024;**12**:100435

[29] Xia Y, Liu J, Chen C, Mo X, Tan Q, He Y, et al. The multifunctions and future prospects of endophytes and their metabolites in plant disease management. Microorganisms. 2022;**10**(5):1072

[30] Khare E, Mishra J, Arora NK. Multifaceted interactions between endophytes and plant: Developments and prospects. Frontiers in Microbiology. 2018;**9**:2732

[31] da Silva JAT, Winarto B, Dobránszki J, Cardoso JC, Zeng S. Tissue disinfection for preparation of culture. Folia Horticulturae. 2016;**28**(1):57-75

[32] Mushtaq S, Shafiq M, Tariq MR, Sami A, Nawaz-ul-Rehman MS, Bhatti MHT, et al. Interaction between bacterial endophytes and host plants. Frontiers in Plant Science. 2023;**13**:1092105

[33] Bahadori Ganjabadi P, Farzaneh M, Mirjalili MH. Development and optimization of culture medium for the production of Glabridin by aspergillus eucalypticola: An endophytic fungus isolated from *Glycyrrhiza glabra* L. (Fabaceae). Mycobiology. 2023;**51**(4):230-238

[34] Dias T, Pimentel V, Cogo AJD, Costa R, Bertolazi AA, Miranda C, et al. The free-living stage growth conditions of the endophytic fungus *Serendipita indica* may regulate its potential as plant growth promoting microbe. Frontiers in Microbiology. 2020;**11**:562238

[35] Andrade GVS, Rodrigues FA, Nadal MC, da Silva Dambroz CM, Martins AD, Rodrigues VA, et al. Plant-endophytic bacteria interactions associated with root and leaf microbiomes of *Cattleya walkeriana* and their effect on plant growth. Scientia Horticulturae. 2023;**309**:111656

[36] Sisti LS, Pena-Passos M, Mayer JLS. Isolation, characterization, and total DNA extraction to identify endophytic fungi in Mycoheterotrophic plants. JoVE (Journal of Visualized Experiments). 2023;**195**:e65135

[37] Matys J, Kensy J, Gedrange T, Zawiślak I, Grzech-Leśniak K, Dobrzyński M. A molecular approach for detecting bacteria and fungi in healthcare environment aerosols: A systematic review. International Journal of Molecular Sciences. 2024;**25**(8):4154

[38] Bielecka M, Pencakowski B, Nicoletti R. Using next-generation sequencing technology to explore genetic pathways in endophytic fungi in the syntheses of plant bioactive metabolites. Agriculture. 2022;**12**(2):187

[39] Fadiji AE, Babalola OO. Metagenomics methods for the study of plant-associated microbial communities: A review. Journal of Microbiological Methods. 2020;**170**:105860

[40] Thomas P, Franco CM. Intracellular bacteria in plants: Elucidation of abundant and diverse cytoplasmic bacteria in healthy plant cells using in vitro cell and callus cultures. Microorganisms. 2021;**9**(2):269

[41] Romano I, Ventorino V, Pepe O. Effectiveness of plant beneficial microbes: Overview of the methodological approaches for the assessment of root colonization and persistence. Frontiers in Plant Science. 2020;**11**:487940

[42] Ahmad T, Farooq S, Mirza DN, Kumar A, Mir RA, Riyaz-Ul-Hassan S. Insights into the endophytic bacterial microbiome of Crocus sativus: Functional characterization leads to potential agents that enhance the plant growth, productivity, and key metabolite content. Microbial Ecology. 2022;**83**(3):669-688

[43] Andrés-Barrao C, Lafi FF, Alam I, de Zélicourt A, Eida AA, Bokhari A, et al. Complete genome sequence analysis of Enterobacter sp. SA187, a plant multi-stress tolerance promoting endophytic bacterium. Frontiers in Microbiology. 2017, 2023:8

[44] Harrison JG, Griffin EA. The diversity and distribution of endophytes across biomes, plant phylogeny and host tissues: How far have we come and where do we go from here? Environmental Microbiology. 2020;**22**(6):2107-2123

[45] Jeger MJ. The impact of climate change on disease in wild plant populations and communities. Plant Pathology. 2022;**71**(1):111-130

[46] Mbareche H, Dumont-Leblond N, Bilodeau GJ, Duchaine C. An overview of bioinformatics tools for DNA meta-barcoding analysis of microbial communities of bioaerosols: Digest for microbiologists. Life. 2020;**10**(9):185

[47] Noguchi M, Toju H. Mycorrhizal and endophytic fungi structure contrasting but interdependent assembly processes in forest below-ground symbiosis. bioRxiv. 2024:1-45. DOI: 10.1101/2024.02.17.580831

[48] Gupta S, Schillaci M, Walker R, Smith PM, Watt M, Roessner U. Alleviation of salinity stress in plants by endophytic plant-fungal symbiosis: Current knowledge, perspectives and future directions. Plant and Soil. 2021;**461**:219-244

[49] Sapkota TB, Takele R. Improving nitrogen use efficiency and reducing nitrogen surplus through best fertilizer nitrogen management in cereal production: The case of India and China. Advances in Agronomy. 2023;**178**:233-294

[50] Eid AM, Fouda A, Abdel-Rahman MA, Salem SS, Elsaied A, Oelmüller R, et al. Harnessing bacterial endophytes for promotion of plant growth and biotechnological applications: An overview. Plants. 2021;**10**(5):935

[51] Rabiey M, Hailey LE, Roy SR, Grenz K, Al-Zadjali MA, Barrett GA, et al. Endophytes vs tree pathogens and pests: Can they be used as biological control agents to improve tree health? European Journal of Plant Pathology. 2019;**155**:711-729

[52] Watts D, Palombo EA, Jaimes Castillo A, Zaferanloo B. Endophytes in agriculture: Potential to improve yields and tolerances of agricultural crops. Microorganisms. 2023;**11**(5):1276

[53] Schulz B, Römmert A-K, Dammann U, Aust H-J, Strack D. The endophyte-host interaction: A balanced antagonism? Mycological Research. 1999;**103**(10):1275-1283

[54] Redman RS, Ranson JC, Rodriguez RJ. Conversion of the pathogenic fungus Colletotrichum magna to a nonpathogenic, endophytic mutualist by gene disruption. Molecular Plant-Microbe Interactions. 1999;**12**(11):969-975

[55] Miller JD, Mackenzie S, Mark F, Adams GW, Findlay JA. Needles of white spruce inoculated with rugulosin-producing endophytes contain rugulosin reducing spruce budworm growth rate. Mycological Research. 2002;**106**(4):471-479

[56] Schulz B, Boyle C, Draeger S, Römmert A-K, Krohn K. Endophytic fungi: A source of novel biologically active secondary metabolites. Mycological Research. 2002;**106**(9):996-1004

[57] Wang W, Barak T, Vinocur B, Shoseyov O, Altman A, editors. Abiotic

resistance and chaperones: Possible physiological role of SP1, a stable and stabilizing protein from Populus. In: Vasil IK, editor. Plant Biotechnology Plant Biotechnology. 2002 and Beyond: Proceedings of the 10 th IAPTC&B Congress June 23-28, 2002 Orlando, Florida, USA. Dordrecht: Springer; 2003

[58] Bray EA. Plant responses to water deficit. Trends in Plant Science. 1997;**2**(2):48-54

[59] Lata R, Chowdhury S, Gond SK, White JF Jr. Induction of abiotic stress tolerance in plants by endophytic microbes. Letters in Applied Microbiology. 2018;**66**(4):268-276

[60] Waqas M, Khan AL, Kamran M, Hamayun M, Kang S-M, Kim Y-H, et al. Endophytic fungi produce gibberellins and indoleacetic acid and promotes host-plant growth during stress. Molecules. 2012;**17**(9):10754-10773

[61] Wani SH, Kumar V, Shriram V, Sah SK. Phytohormones and their metabolic engineering for abiotic stress tolerance in crop plants. The Crop Journal. 2016;**4**(3):162-176

[62] Chitnis VR, Suryanarayanan TS, Nataraja KN, Prasad SR, Oelmüller R, Shaanker RU. Fungal endophyte-mediated crop improvement: The way ahead. Frontiers in Plant Science. 2020;**11**:561007

[63] Ahmad MF, Alsayegh AA, Ahmad FA, Akhtar MS, Alavudeen SS, Bantun F, et al. Ganoderma lucidum: Insight into antimicrobial and antioxidant properties with development of secondary metabolites. Heliyon. 2024;**10**:1-19

[64] Priyashantha A, Karunarathna S, Lu L, Tibpromma S. Fungal endophytes: An alternative biocontrol agent against phytopathogenic fungi. Encyclopedia. 2023;**2023**(3):759-780

[65] Ayivi RD, Obare SO, Wei J. Molecularly imprinted polymers as chemosensors for organophosphate pesticide detection and environmental applications. TrAC Trends in Analytical Chemistry. 2023;**117231**

[66] Lechenet M, Dessaint F, Py G, Makowski D, Munier-Jolain N. Reducing pesticide use while preserving crop productivity and profitability on arable farms. Nature Plants. 2017;**3**(3):1-6

[67] Gupta I, Singh R, Muthusamy S, Sharma M, Grewal K, Singh HP, et al. Plant essential oils as biopesticides: Applications, mechanisms, innovations, and constraints. Plants. 2023;**12**(16):2916

[68] Mesguida O, Haidar R, Yacoub A, Dreux-Zigha A, Berthon J-Y, Guyoneaud R, et al. Microbial biological control of fungi associated with grapevine trunk diseases: A review of strain diversity, modes of action, and advantages and limits of current strategies. Journal of Fungi. 2023;**9**(6):638

[69] Dong L, Huang Z. Some evidence and new insights for feedback loops of human-nature interactions from a holistic earth perspective. Journal of Cleaner Production. 2023;**139667**:1-15

[70] Mhlongo MI, Piater LA, Madala NE, Labuschagne N, Dubery IA. The chemistry of plant–microbe interactions in the rhizosphere and the potential for metabolomics to reveal signaling related to defense priming and induced systemic resistance. Frontiers in Plant Science. 2018;**9**:112

[71] Pérez-Montaño F, Alías-Villegas C, Bellogín R, Del Cerro P, Espuny M, Jiménez-Guerrero I, et al. Plant growth promotion in cereal and leguminous agricultural important plants: From microorganism capacities to crop production. Microbiological Research. 2014;**169**(5-6):325-336

[72] Kashyap N, Singh SK, Yadav N, Singh VK, Kumari M, Kumar D, et al. Biocontrol screening of endophytes: Applications and limitations. Plants. 2023;**12**(13):2480

[73] Jaffar NS, Jawan R, Chong KP. The potential of lactic acid bacteria in mediating the control of plant diseases and plant growth stimulation in crop production-A mini review. Frontiers in Plant Science. 2023;**13**:1047945

[74] Latz MA, Jensen B, Collinge DB, Jørgensen HJ. Endophytic fungi as biocontrol agents: Elucidating mechanisms in disease suppression. Plant Ecology & Diversity. 2018;**11**(5-6):555-567

[75] Hasan A, Tabassum B, Hashim M, Khan N. Role of plant growth promoting rhizobacteria (PGPR) as a plant growth enhancer for sustainable agriculture: A review. Bacteria. 2024;**3**(2):59-75

[76] Akram S, Ahmed A, He P, He P, Liu Y, Wu Y, et al. Uniting the role of endophytic fungi against plant pathogens and their interaction. Journal of Fungi. 2023;**9**(1):72

[77] Ren N, Wang L, You C. Diversity, community structure, and antagonism of endophytic fungi from asymptomatic and symptomatic Mongolian pine trees. Journal of Fungi. 2024;**10**(3):212

[78] Patel JK, Mistry Y, Soni R, Jha A. Evaluation of antifungal activity of endophytic Bacillus spp. and identification of secondary metabolites produced against the phytopathogenic fungi. Current Microbiology. 2024;**81**(5):128

[79] Muhammad M, Basit A, Ali K, Ahmad H, Li W-j, Khan A, et al. A review on endophytic fungi: A potent reservoir of bioactive metabolites with special emphasis on blight disease management. Archives of Microbiology. 2024;**206**(3):129

[80] Fontana DC, de Paula S, Torres AG, de Souza VHM, Pascholati SF, Schmidt D, et al. Endophytic fungi: Biological control and induced resistance to phytopathogens and abiotic stresses. Pathogens. 2021;**10**(5):570

[81] Yu Y, Gui Y, Li Z, Jiang C, Guo J, Niu D. Induced systemic resistance for improving plant immunity by beneficial microbes. Plants. 2022;**11**(3):386

[82] Ghosh S, Bhagwat T, Webster TJ. Endophytic microbiomes and their plant growth-promoting attributes for plant health. Current Trends in Microbial Biotechnology for Sustainable Agriculture. 2021. pp. 245-278. ISBN: 978-981-15-6949-4

[83] Khan N, Ali S, Shahid MA, Mustafa A, Sayyed R, Curá JA. Insights into the interactions among roots, rhizosphere, and rhizobacteria for improving plant growth and tolerance to abiotic stresses: A review. Cell. 2021;**10**(6):1551

[84] Chaudhary P, Agri U, Chaudhary A, Kumar A, Kumar G. Endophytes and their potential in biotic stress management and crop production. Frontiers in Microbiology. 2022;**13**:933017

[85] Pandit MA, Kumar J, Gulati S, Bhandari N, Mehta P, Katyal R, et al. Major biological control strategies for plant pathogens. Pathogens. 2022;**11**(2):273

[86] Morales-Cedeño LR, del Carmen O-MM, Loeza-Lara PD, Parra-Cota FI, de Los S-VS, Santoyo G. Plant growth-promoting bacterial endophytes as biocontrol agents of pre-and post-harvest diseases: Fundamentals, methods of application and future perspectives. Microbiological Research. 2021;**242**:126612

[87] Nadarajah K, Abdul Rahman NSN. The microbial connection to sustainable agriculture. Plants. 2023;**12**(12):2307

[88] Roca-Couso R, Flores-Félix JD, Rivas R. Mechanisms of action of microbial biocontrol agents against Botrytis cinerea. Journal of Fungi. 2021;**7**(12):1045

[89] Aswani R, Thomas R, Radhakrishnan E. Induction of plant defense response by endophytic microorganisms. In: Biocontrol Mechanisms of Endophytic Microorganisms. Acadmic Press, Elsevier; 2022. pp. 89-115

[90] Hossain MT, Akhter MS, Islam MM, Faruk MI, Chung YR. Cross-talks about hemibiotrophic-necrotrophic pathogens by endophytic Bacillus-based EMOs. In: Plant Endophytes and Secondary Metabolites. Acadmic Press, Elsevier; 2024. pp. 235-253

[91] Wippel K. Plant and microbial features governing an endophytic lifestyle. Current Opinion in Plant Biology. 2023;**76**:102483

[92] Wang Y-L, Zhang H-B. Assembly and function of seed endophytes in response to environmental stress. Journal of Microbiology and Biotechnology. 2023;**33**(9):1119

[93] Lu H, Wei T, Lou H, Shu X, Chen Q. A critical review on communication mechanism within plant-endophytic fungi interactions to cope with biotic and abiotic stresses. Journal of Fungi. 2021;**7**(9):719

[94] Gjennestad RS. A Study of PAMP-INDUCED SECRETED PEPTIDE LIKE 6 (PIPL6) in Regulation of Plant Immunity. NTNU; 2021

[95] Pinski A, Betekhtin A, Hupert-Kocurek K, Mur LA, Hasterok R. Defining the genetic basis of plant–endophytic bacteria interactions. International Journal of Molecular Sciences. 2019;**20**(8):1947

[96] Fadiji AE, Babalola OO. Elucidating mechanisms of endophytes used in plant protection and other bioactivities with multifunctional prospects. Frontiers in Bioengineering and Biotechnology. 2020;**8**:467

[97] Medison RG, Tan L, Medison MB, Chiwina KE. Use of beneficial bacterial endophytes: A practical strategy to achieve sustainable agriculture. AIMS Microbiology. 2022;**8**(4):624

[98] Sahu PK, Jayalakshmi K, Tilgam J, Gupta A, Nagaraju Y, Kumar A, et al. ROS generated from biotic stress: Effects on plants and alleviation by endophytic microbes. Frontiers in Plant Science. 2022;**13**:1042936

[99] Phurailatpam L, Mishra S. Role of plant endophytes in conferring abiotic stress tolerance. In: Plant Ecophysiology and Adaptation under Climate Change: Mechanisms and Perspectives II: Mechanisms of Adaptation and Stress Amelioration. Singapore: Springer; 2020. pp. 603-628

[100] Mathur P, Roy S. Insights into the plant responses to drought and decoding the potential of root associated microbiome for inducing drought tolerance. Physiologia Plantarum. 2021;**172**(2):1016-1029

[101] Shaffique S, Khan MA, Wani SH, Pande A, Imran M, Kang S-M, et al. A review on the role of endophytes and plant growth promoting rhizobacteria in mitigating heat stress in plants. Microorganisms. 2022;**10**(7):1286

[102] Khan N, Bano A, Ali S, Babar MA. Crosstalk amongst phytohormones from planta and PGPR under biotic and abiotic

stresses. Plant Growth Regulation. 2020;**90**:189-203

[103] Jin R, Kim BH, Ji CY, Kim HS, Li HM, Kwak S-S. Overexpressing IbCBF3 increases low temperature and drought stress tolerance in transgenic sweetpotato. Plant Physiology and Biochemistry. 2017;**118**:45-54

[104] Jia X, Qi E, Liu S, Ma S, Wen G, Zhang X, et al. Physiological and transcriptomic analysis reveals drought-stress responses of Arabidopsis DREB1A in transgenic potato. Potato Research. 2023;**66**(4):1143-1164

[105] Kaya C, Uğurlar F, Adamakis I-DS. Epigenetic and hormonal modulation in plant–plant growth-promoting microorganism symbiosis for drought-resilient agriculture. International Journal of Molecular Sciences. 2023;**24**(22):16064

[106] Afreen U, Kumar M, Mukhopadhyay K. Small RNA networking: Host-microbe interaction in food crops. In: Plant Small RNA in Food Crops. Elsevier, Acadmic Press; 2023. pp. 271-293

[107] Mengistu AA. Endophytes: Colonization, behaviour, and their role in defense mechanism. International Journal of Microbiology. 2020;**2020**:1-8

[108] Pandey PK, Samanta R, Yadav RNS. Inside the plant: Addressing bacterial endophytes in biotic stress alleviation. Archives of Microbiology. 2019;**201**:415-429

[109] Yang Y-X, Ahammed GJ, Wu C, Fan S-Y, Zhou Y-H. Crosstalk among jasmonate, salicylate and ethylene signaling pathways in plant disease and immune responses. Current Protein and Peptide Science. 2015;**16**(5):450-461

[110] Oukala N, Aissat K, Pastor V. Bacterial endophytes: The hidden actor in plant immune responses against biotic stress. Plants. 2021;**10**(5):1012

[111] Lopes NS, Santos AS, de Novais DPS, Pirovani CP, Micheli F. Pathogenesis-related protein 10 in resistance to biotic stress: Progress in elucidating functions, regulation and modes of action. Frontiers in Plant Science. 2023;**14**:1193873

[112] Sood M, Kapoor D, Kumar V, Kalia N, Bhardwaj R, Sidhu GP, et al. Mechanisms of plant defense under pathogen stress: A review. Current Protein and Peptide Science. 2021;**22**(5):376-395

[113] Kumar V, Nautiyal CS. Endophytes modulate plant genes: Present status and future perspectives. Current Microbiology. 2023;**80**(11):353

[114] Lephatsi MM, Meyer V, Piater LA, Dubery IA, Tugizimana F. Plant responses to abiotic stresses and rhizobacterial biostimulants: Metabolomics and epigenetics perspectives. Metabolites. 2021;**11**(7):457

[115] Senanayake I, Rathnayaka A, Marasinghe D, Calabon M, Gentekaki E, Lee H, et al. Morphological approaches in studying fungi: Collection, examination, isolation, sporulation and preservation. Mycosphere. 2020;**11**(1):2678-2754

[116] Rodriguez R, White J Jr, Arnold A, Redman RS. Fungal endophytes: Diversity and functional roles. New Phytologist. 2009;**182**(2):314-330

[117] Omomowo OI, Babalola OO. Bacterial and fungal endophytes: Tiny giants with immense beneficial potential for plant growth and sustainable agricultural productivity. Microorganisms. 2019;**7**(11):481

[118] Hassanisaadi M, Barani M, Rahdar A, Heidary M, Thysiadou A, Kyzas GZ. Role of agrochemical-based nanomaterials in plants: Biotic and abiotic stress with germination improvement of seeds. Plant Growth Regulation. 2022;**97**(2):375-418

[119] Dubey A, Malla MA, Kumar A, Dayanandan S, Khan ML. Plants endophytes: Unveiling hidden agenda for bioprospecting toward sustainable agriculture. Critical Reviews in Biotechnology. 2020;**40**(8):1210-1231